SpringerBriefs in Computer Science

SpringerBriefs present concise summaries of cutting-edge research and practical applications across a wide spectrum of fields. Featuring compact volumes of 50 to 125 pages, the series covers a range of content from professional to academic.

Typical topics might include:

- A timely report of state-of-the art analytical techniques
- A bridge between new research results, as published in journal articles, and a contextual literature review
- A snapshot of a hot or emerging topic
- An in-depth case study or clinical example
- A presentation of core concepts that students must understand in order to make independent contributions

Briefs allow authors to present their ideas and readers to absorb them with minimal time investment. Briefs will be published as part of Springer's eBook collection, with millions of users worldwide. In addition, Briefs will be available for individual print and electronic purchase. Briefs are characterized by fast, global electronic dissemination, standard publishing contracts, easy-to-use manuscript preparation and formatting guidelines, and expedited production schedules. We aim for publication 8–12 weeks after acceptance. Both solicited and unsolicited manuscripts are considered for publication in this series.

**Indexing: This series is indexed in Scopus, Ei-Compendex, and zbMATH **

Lu Hou • Lingyi Han • Kan Zheng

Blockchain-Based Internet of Things

 Springer

Lu Hou
Beijing University of Posts
and Telecommunications
Beijing, China

Lingyi Han
China Aero Geophysical Survey & Remote
Sensing Center for Natural Resources
Beijing, China

Kan Zheng
Ningbo University
Zhejiang, China

ISSN 2191-5768   ISSN 2191-5776   (electronic)
SpringerBriefs in Computer Science
ISBN 978-3-031-70302-7   ISBN 978-3-031-70303-4   (eBook)
https://doi.org/10.1007/978-3-031-70303-4

© The Editor(s) (if applicable) and The Author(s), under exclusive license to Springer Nature Switzerland AG 2024

This work is subject to copyright. All rights are solely and exclusively licensed by the Publisher, whether the whole or part of the material is concerned, specifically the rights of translation, reprinting, reuse of illustrations, recitation, broadcasting, reproduction on microfilms or in any other physical way, and transmission or information storage and retrieval, electronic adaptation, computer software, or by similar or dissimilar methodology now known or hereafter developed.
The use of general descriptive names, registered names, trademarks, service marks, etc. in this publication does not imply, even in the absence of a specific statement, that such names are exempt from the relevant protective laws and regulations and therefore free for general use.
The publisher, the authors and the editors are safe to assume that the advice and information in this book are believed to be true and accurate at the date of publication. Neither the publisher nor the authors or the editors give a warranty, expressed or implied, with respect to the material contained herein or for any errors or omissions that may have been made. The publisher remains neutral with regard to jurisdictional claims in published maps and institutional affiliations.

This Springer imprint is published by the registered company Springer Nature Switzerland AG
The registered company address is: Gewerbestrasse 11, 6330 Cham, Switzerland

If disposing of this product, please recycle the paper.

Preface

This book specifically focuses on incorporating blockchain technology into Internet of Things (IoT) systems, delving into the methodologies used for each element of the system. We will discuss the essential tactics required to deploy blockchain in order to improve the security and privacy of IoT devices. As the development of IoT applications progresses, the data produced by IoT systems demonstrates substantial commercial worth. Nevertheless, this also renders security vulnerabilities and privacy concerns critical as a result of the absence of essential security safeguards. Blockchain, as a potential solution for decentralized storage and consensus, has a substantial influence on data security and privacy. Decentralized nodes collaborate to create a blockchain network, where each node has a full copy of the ledger. Data is transmitted by transactions and stored in chained blocks using a consensus mechanism. Furthermore, smart contracts enable the functioning of blockchain applications and construction of trust among participants. The blockchain is well-suited for enhancing data security and privacy in the context of the IoT. Nevertheless, it introduces additional overhead that reduces the efficiency of the IoT system. Hence, it is imperative to meticulously build transaction and block structures tailored for IoT data. Additionally, the process of generating blocks and the decentralized consensus mechanism must be efficient and lightweight, as IoT devices have limited resources. Overall, this book provides multiple Artificial Intelligence (AI)-based schemes to tackle these challenges. This book is appropriate for academics and college students studying relevant topics such as the IoT, blockchain, wireless communications, and AI.

We are very grateful to Prof. Xuemin (Sherman) Shen, the SpringerBriefs Series Editor on Wireless Communications. This book would not be possible without his kind support. Special thanks are also attributed to Priyadarsini K. and Susan Lagerstrom-Fife at Springer Nature, for their assistance throughout the preparation process of this monograph.

We would like to thank Zhiming Liu and Qihao Zhou from the Intelligent Communication & Computing (IC^2) Group at the Beijing University of Posts & Telecommunications (BUPT) for their contributions to the work presented in this monograph. We also would like to thank all the members of the IC^2 group for their valuable discussions and insightful suggestions, ideas, and comments.

This work is funded by the National Science Foundation of China (No. 62001052).

Beijing, China   Lu Hou
Beijing, China   Lingyi Han
Zhejiang, China   Kan Zheng

Contents

Acronyms

AI	Artificial Intelligence
BIoT	Blockchain-based Internet of Things
CFT	Crash Fault Tolerance
DAG	Direct-Acyclic-Graph
DDPG	Deep Deterministic Policy Gradient
DoS	Denial of Services
DQN	Deep Q Network
IoT	Internet of Things
IIoT	Industrial Internet of Things
JS	Join Server
LoRa	Long Range
LPWA	Low-Power Wide-Area
LTE	Long-Term Evolution
MDP	Markov Decision Process
MEC	Multi-access Edge Computing
MSP	Membership Service Provider
MTC	Machine-Type Communication
NB-IoT	Narrow Band IoT
NC	Network Controller
PBFT	Practical Byzantine Fault Tolerance
PoS	Proof of Stake
PoW	Proof of Work
SNR	Signal-to-Noise Radio
SoC	System on Chip
SPOF	Single Point of Failure
URLLC	Ultra-Reliable Low-Latency Communication
5G	Fifth Generation

Chapter 1
Introduction

With the rapid development of the IoT, security and privacy issues have garnered significant attention from both industry and academia. The commercial value of IoT data is increasing alongside the expansion of IoT applications, leading to rising demands for IoT data security and privacy. Ensuring IoT data security involves maintaining data immutability and disaster tolerance, while privacy requires that data access is restricted to authorized users only. Blockchain technology is emerging as a promising solution for providing transparent and secure data storage and authentication for IoT applications [1]. However, directly integrating blockchain into IoT systems poses several challenges that need to be addressed. This monograph aims to explore the key challenges associated with Blockchain-based Internet of Things (BIoT). We intend to provide a feasible and efficient BIoT solution to meet the data security demands of IoT. In this chapter, we first discuss the fundamentals of BIoT and the motivations for using blockchain in IoT. Next, we examine the specific challenges of BIoT integration. Finally, we present the main subject of the monograph.

1.1 Internet of Things

The advancement of communication, information, and computer science has resulted in the extensive utilization of IoT. The fundamental principle of an IoT system is the collection of diverse data from sensors using wired/wireless communication links, which is then transmitted to servers. These servers subsequently provide control signals to the actuators.

© The Author(s), under exclusive license to Springer Nature Switzerland AG 2024
L. Hou et al., *Blockchain-Based Internet of Things*, SpringerBriefs in Computer Science, https://doi.org/10.1007/978-3-031-70303-4_1

1.1.1   Communications in IoT

Wireless communication technologies are essential for the continuing growth and advancement of the IoT, driving its continuous development and expansion. The high speed, low latency, and high capacity of Fifth Generation (5G) systems make it an optimal option for connecting a vast number of IoT devices. 5G enables IoT devices to reach enhanced data transmission speeds and more connections, hence facilitating many application scenarios such as smart cities, intelligent transportation, and industrial automation.

Second, Long Range (LoRa), as one of the Low-Power Wide-Area (LPWA) technologies, also plays an important role in IoT. LoRa features long-distance communication and low power consumption, making it suitable for applications requiring low cost and low power consumption, such as remote monitoring, environmental sensing, and agriculture.

Additionally, Wi-Fi technology, as a common wireless local area network technology, is widely used in home and enterprise environments to connect various IoT devices. Wi-Fi provides high-speed LAN connections, allowing users to conveniently access and manage various smart devices, such as smart home devices, smart cameras, and smart speakers.

Thus, wireless communication technologies such as 5G, LoRa, and Wi-Fi each have their own characteristics for the IoT, collectively promoting its development and providing reliable connection and communication support for various applications.

- **5G**

  By 2025, IoT devices are projected to produce over 50 % half of the global data. Transmitting such data and fully harnessing its capabilities would require 5G wireless bandwidth to be 1000 times greater than the capabilities of Long-Term Evolution (LTE).

  Ultra-Reliable Low-Latency Communication (URLLC) in 5G enables the deployment of safety-critical applications such as infrastructure or manufacturing systems. The massive Machine-Type Communication (MTC) provides a large volume of connections from IoT devices, while the Narrow Band IoT (NB-IoT) offers low-power communication capabilities. Besides, the utilization of 5G technology will expedite edge computing, thereby bringing computational capabilities closer to end-users. This will enable the provision of real-time intelligence while ensuring that the data remain localized. Overall, 5G plays a crucial role in enabling IoT applications.

- **LoRa**

  LoRa is anticipated to offer energy-efficient communications to a vast number of end-devices in a wide area [2]. LoRa is appealing for a range of IoT applications, including environment monitoring and Industrial Internet of Things (IIoT) [3]. LoRa gateways can gather a significant volume of IoT data with a low sampling rate. These data can then be processed by the central cloud [4]. Nevertheless, there are several obstacles that must be overcome prior to the widespread implementation of LoRa networks. The majority of them pertain to enhancing the efficacy and safeguarding of the LoRa systems.

The existing LoRa gateways have the only responsibility of transparently forwarding packages between end-devices and the central cloud. Subsequently, the central cloud is responsible for managing and storing all LoRa packages and undertakes all of the workloads [5]. Therefore, the computational and storage resources of the LoRa gateways are underutilized. Furthermore, the central cloud is located at a considerable distance from the end-devices, which poses a challenge in meeting the low-latency requirements of time-sensitive IoT applications [6]. LoRa systems also provide possible security problems. For instance, the LoRa data are susceptible to forgery and destruction, as they are all collected in a central cloud [7, 8]. For a comprehensive explanation of LoRa, please refer to Sect. 5.3.

1.1.2   Architecture of IoT System

1.1.2.1   Basic Architecture

The fundamental architecture of an IoT system is illustrated in Fig. 1.1. The IoT system consists of several essential components, including IoT servers, base stations, IoT gateways, and end-devices. The end-devices, comprising sensors and actuators, are of utmost importance in this architecture. Sensors consistently monitor the physical environment and upload the collected data to the IoT servers through gateways or base stations. The data gathered by IoT servers are analyzed and used to generate commands that are subsequently transmitted to the end-devices. These

Fig. 1.1 Illustration of a typical IoT system

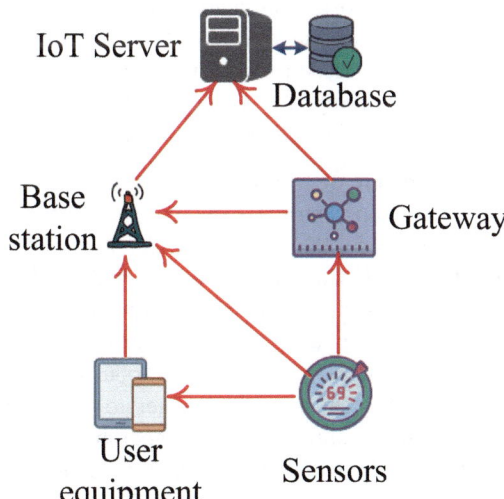

commands allow for various activities to be performed, such as controlling switches, valves, or robots. All data generated within the system are stored in databases.

1.1.2.2 Edge Computing

Edge computing enables the localization of data processing in close proximity to IoT devices. These benefits can include improved latency, enhanced performance, reduced costs, and increased security advantages for enterprise IT. Instead of transmitting data to be analyzed on remote cloud servers or centralized data centers, which consumes valuable time and resources, the computation occurs directly on the device or within the network. Subsequently, the processed data can be expedited to its intended location. Edge computing mitigates the series of possible limitations in data transmission capacity and handles the relevant data by retaining it in close to its source.

The edge computing servers act as an intermediary layer between end-devices and cloud servers. Certain tasks performed by the central cloud can be offloaded to gateways or base stations, including the management of end-devices and the fundamental processing of data packages. In order to enhance the quality of services, it is imperative to fully utilize the resources available at the network edge. Furthermore, it is imperative for the centralized paradigm to transition into a decentralized one.

1.1.3   Challenges of IoT System

IoT systems typically employ a centralized architecture, in which the essential services are hosted on central clusters. Moreover, data from every geographical locations are subjected to processing and subsequently stored in cloud servers. The centralized service method offers the benefit of less administrative costs, as all cloud servers are deployed in a single location. However, these centralized systems have certain limitations that must be resolved:

- **Significant demand for high server capability**. In order to operate all the services on centralized servers and account for potential future scalability, it is necessary for the centralized servers to possess a high level of competence.
- **High cost**. Centralized servers with high capability and scalability demands necessitate significant capital expenditures. As the emerging of new applications and data, the business value of data increases. Therefore, the central cloud needs to have greater capability and robustness. For example, it is necessary to lease more physical servers in order to guarantee a high level of data processing capacity. Additionally, it is important to distribute these physical servers across data centers situated in different geographical areas as a backup measure.

- **High risk of single-node failure**. If the central node malfunctions, the entire system will become inaccessible. Conversely, all the running stresses are also consolidated. The central servers have the responsibility of gathering, storing, analyzing, and processing all data across all domains. Simultaneously, they must oversee all the end-devices throughout all regions. End-devices only require the ability to sense the data or be controlled by central commands. Base stations or gateways simply facilitate the seamless transmission of IoT data between end-devices and central servers. Consequently, there is a significant disparity in the utilization of computational resources between central servers and edge servers. The majority of computing tasks are conducted by central servers, resulting in the squandering of resources in base stations, gateways, and end-devices.
- **Low efficiency**. The collection of all IoT data necessitates the utilization of sensors, which then transmit the data through a series of components including a gateway, base station, core network, and the Internet, ultimately reaching the centralized servers. The latency of these data is prolonged, resulting in a decrease in its freshness. Furthermore, end-users are required to engage with end-devices through the central servers, resulting in a significant delay in response time and a subpar quality of services.
- **Absolute trust of service provider**. Complete reliance on centralized services providers is necessary for all participants, as these providers have exclusive possession of all IoT data. The establishment of trust relies on subjective faith, societal norms, and legal frameworks, rather than being dependent on technology. In other words, the data creator lacks the technical means to prohibit unauthorized access to the data by service providers.
- **Data leakage**. The entirety of IoT data is transmitted to the central servers, presenting significant challenges in terms of data security and privacy. The data may be intercepted by attackers, while they are transmitted over the Internet. On the other hand, there may exist unauthorized or unintentional release of sensitive or confidential information. The data leakage can cause huge economic losses.

1.2 Blockchain

1.2.1 Hash Function

Blockchain is a specific type of data structure that utilizes hash operations to link blocks together in a sequential manner, as depicted in Fig. 1.2. Typically, the hash function possesses the subsequent traits:

- **Hiding**: The hash function can *hide* the original input. In other words, if we have an output value y from a hash function $\mathcal{H}$, it is not possible to derive the input x that satisfies $y = \mathcal{H}(x)$ without using brute force over all possible values of x.
- **Collision resistance**: Given any two input values to be x and y, if $x \neq y$ holds, then the output values of the hash function satisfy $\mathcal{H}(x) \neq \mathcal{H}(y)$. It also implies

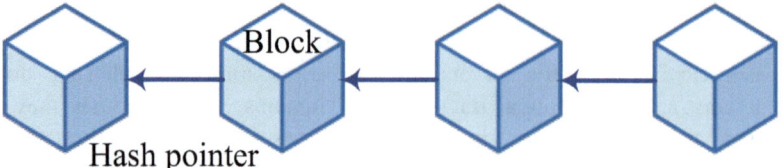

Fig. 1.2 Illustration of the blockchain structure

that making merely a minimal adjustment to the input value x will result in a distinct output value y.

- **Puzzle friendly**: The puzzle entails identifying the precise value of x that corresponds to the provided hash value. According to the *Hiding* property, the output value y does not reveal any information about the input value. However, if the input value x is known, it can be easily verified that it is indeed the input value that produces the output value y by the hash function. That is to say, the puzzle is difficult to solve, but easy to verify.

The hash function is extensively utilized in blockchain technology to maintain immutability of the data, as elaborated in the following subsections.

1.2.2 Block Structure

A block serves as the fundamental unit in a blockchain system. Every block is comprised of a header and a body. The primary characteristic of the blockchain is the unchangeable nature of the ledger. This is realized by a particular field of the block header, namely parent block hash. The parent block hash refers to the hash value of the *previous block header*. By chaining all the blocks together, even a small adjustment to the content of a block will cause the *parent block hash* field of the following block to change, as well as the *parent block hash* field of the subsequent block, all the way up to the most recent block. Various types of blockchains employ different organizational systems for content of the block body, but they all necessitate the inclusion of a hash value in the block header. Therefore, any slight alteration to the block body results in substantial revisions to the block headers, which extend all the way to the most recent block. These inherent characteristics are the reason why blockchain is deemed immutable.

1.2.3 Transaction

The main purpose function of blockchains like Bitcoin and Ethereum is to facilitate the exchange of digital currency. The transfer is executed by initiating a transaction from sender to receiver. Transactions on Ethereum encompass the responsibility of both deploying and executing smart contracts. The individual with the authority to authenticate creates a fresh block in which the block content consists of these transactions, who has the right to account, and constructs a new block where the block body is made up of these transactions. Blockchain uses hash-based tree structures, such as Merkle tree in Bitcoin, Merkle Partricia tree in Ethereum or Merkle Radix Tree in other platforms, to organize transactions in each block, and a transaction root hash value is stored in block header for immutability.

1.2.4 Blockchain Category

Blockchains can be categorized based on the participants involved, such as public blockchain, private blockchain, and the consortium blockchain. The categories and characteristics of each type of blockchain are summarized in Table 1.1.

1. **Public blockchain**

   Public blockchains are accessible to the general public through the Internet. Users have the freedom to join or quit the public blockchains at their own discretion. Simultaneously, all the data stored on the blockchain are made public. The primary objective of a public blockchain is to establish trust on the Internet and offer a decentralized and secure method of transferring currency. The public blockchain provides users with the guarantee that the data stored on the chain is authentic and immutable. Out of these public blockchains, two stand out as they were the first to implement certain features of blockchain.

   (a) **Bitcoin** was designed and put into practice by Satoshi Nakamoto in 2008. The blockchain technology is being unveiled globally for the first time. The Bitcoin network is upheld by a collection of nodes referred to as miners.

Table 1.1 Blockchain categories and characteristics [8]

Category	Public blockchain	Private blockchain	Consortium blockchain
Decentralization	Fully	Centralized	Partial
Consensus	PoW, PoS	Ripple	PBFT, RAFT
Permission	Permissionless	Permission	Permission
Representative	Bitcoin, Ethereum	GemOS	Hyperledger
Smart contract	Solidity	–	Chaincode
Advantage	High security	High flexibility	High efficiency
Disadvantage	High latency, Low throughput	Trustless	Limited participants
Application	Digital currency, Decentralized finance	Intra-company audit	Digital business, Company cooperation

Every miner possesses a ledger, often known as a blockchain, and achieves consensus on the blockchain's content by solving a mining puzzle referred to as Proof of Work (PoW).

(b) **Ethereum** was founded by Vitalik Buterin in 2013. The introduction of the smart contract concept has led to the recognition of this as the blockchain version 2.0. Every node in Ethereum operates an Ethereum virtual machine that enables the execution of intricate programs written in a Turing-complete language. Consequently, a decentralized platform has the capacity to host numerous applications, prompting us to consider the new approach to manage data in the IoT.

2. **Private blockchain**

The private blockchain is fully contained within an enterprise. It is under the control of a sole entity. Hence, the decentralized aspect of blockchain serves no use. A centralized institution can utilize conventional databases to ensure the protection of private data.

3. **Consortium blockchain** Consortium blockchain is another commonly utilized form of blockchain. Unlike the public blockchain, a consortium blockchain only permits access to authorized entities. A consortium blockchain employs a membership service provider to oversee the participation of members. The reading and writing privileges are determined and governed by all the members involved. The primary objective of a consortium blockchain is to achieve consensus and establish trust among a selected group of participants.

Consortium blockchains are particularly suitable for IoT systems due to their ability to facilitate private data transit and storage [9].

1.2.5 Features of Blockchain

The characteristics of blockchain can be succinctly summarized as follows:

- **Decentralization:** In a typical market system, the verification of transactions and the management of identities are entrusted to reliable third-party organizations or companies, such as the bank that issues credit cards. The credit degree is also dependent on third parties. Blockchain enables peer-to-peer trading between two entities of equal standing, with the verification, storage, and consensus on transactions being managed by participants who have separate owners. Thus, a completely autonomous system is built, in which no centralized components have the ability to assume control over the system.
- **Immutability:** As explained in Sect. 1.2.2, the use of a chain structure with hash pointers guarantees that all the data on the blockchain cannot be changed. Any

alteration leads to the alteration of the hash value of the current block, as well as the hash value of the following block, and so on. Due to the decentralized nature of blockchain, obtaining approval from all other nodes for these changes is exceedingly challenging to achieve.

- **Non-repudiation:** All transactions necessitate digital signatures from the individuals initiating them. The digital signatures are created using the private keys held by the initiators, and those possessing public keys can authenticate the signatures. Thus, the initiator cannot deny that the transaction is sent by it.
- **Transparency:** In the majority of public blockchains, every node has the ability to access all the data stored on the blockchain. In other words, all the information stored on blockchain is easily visible and accessible to everybody.
- **Anonymity:** While the information stored on the blockchain is visible to anyone, a participant can initiate a transaction using only an address. The address is determined using a hash function, which is independent of the actual identity of the participant. Other validators are limited to obtaining information solely about the transaction addresses of the sender and receiver.
- **Traceability:** The blockchain's immutability and non-repudiation features enable the permanent recording of any user's actions on the blockchain. These records are immutable and tamper-proof and can be authenticated by all validators. Hence, the activities involving data on a blockchain, such as creation, alteration, retrieval, and deletion, can be easily tracked.

1.3   Necessity of Blockchain-Based IoT

1.3.1   Limited Capability

The blockchain system necessitates that all participants maintain identical ledgers, verify transactions, and reach consensus. Except for IoT servers, the other entities have restricted capacity in terms of computation, storage, and energy, particularly end-devices. End-devices have limited storage capacities in order to maintain tiny sizes. Consequently, it is difficult for end-devices to connect to the blockchain network. The ledgers must be stored in base stations or gateways in order to utilize the edge capability. Ensuring the functionality of a standard blockchain system at the edge necessitates a substantial quantity of computational resources for validating transactions and executing consensus methods, as well as ample storage capacity for housing the distributed ledgers. Hence, while implementing blockchain for an IoT system with edge computing, it is crucial to account for the restricted capacity. The blockchain must possess a low weight and occupy little space.

1.3.2   Latency and Throughput

Certain IoT applications necessitate immediate and instantaneous reactions. While it is possible to implement the blockchain at the network edge, which is in close proximity to the user, many processes involved in the blockchain, such as initiating transactions, constructing blocks, executing smart contracts, and reaching consensus, require a significant amount of time and are limited in terms of throughput. This challenge undermines the implementation of blockchain in IoT. Hence, in order to construct a blockchain-based IoT system, it is imperative to enhance the latency and throughput of applications.

1.3.3   Implementation Feasibility

Is it conceivable to execute the concept of integrating blockchain with IoT? The platforms utilized for IoT systems and blockchain systems are distinct. As an illustration, IoT gateways are constructed using embedded devices, whereas regular blockchain systems operate on personal computers, workstations, or high-performance servers. These platforms possess distinct architectures and operation systems. Therefore, the viability of a blockchain-based IoT system is ambiguous.

1.4   Aim of the Monograph

As previously discussed, the BIoT system has the capability to provide robust data security and privacy, while also establishing a reliable platform for many parties. The expenses associated with BIoT include the increased workloads on IoT entities with limited resources, the incremented delay in IoT applications, and the decrease in throughput. Therefore, this monograph seeks to present a practical architecture of BIoT system that effectively tackles the aforementioned obstacles. Our monograph provides a detailed analysis of the transaction and block structure, as well as the consensus mechanism, used in the BIoT system. Additionally, we explore various optimization strategies that aim to minimize latency and resource usage. Furthermore, we provide a practical demonstration of out BIoT concept by implementing a prototype on a genuine IoT system, thereby showcasing its viability.

The rest of the monograph is organized as follows. An architecture design is provided in Chap. 2, which illustrates how to build up a BIoT system, including the ledger structure, transaction and block forms, and consensus mechanism. Chapter 3 provides an intelligent transaction migration scheme in order to balance the burden of different gateways on new block generation. In Chap. 4, we present a consensus mechanism for BIoT system to enhance the efficiency and meet the requirements of IoT applications. We also give a prototype implementation that utilizes the consortium blockchain in the LoRa system in Chap. 5. Finally, we provide conclusions and outlooks to guide future researchers in the field of blockchain for IoT applications in Chap. 6.

References

1. C.H. Liu, Q. Lin, S. Wen, Blockchain-enabled data collection and sharing for industrial IoT with deep reinforcement learning. IEEE Trans. Ind. Inform. **15**(6), 3516–3526 (2019)
2. X. Xiong, K. Zheng, R. Xu, W. Xiang, P. Chatzimisios, Low power wide area machine-to-machine networks: key techniques and prototype. IEEE Commun. Mag. **53**(9), 64–71 (2015)
3. L. Leonardi, F. Battaglia, L. Lo Bello, RT-LoRa: a medium access strategy to support real-time flows over LoRa-based networks for industrial IoT applications. IEEE Internet Things J. **6**(6), 10 812–10 823 (2019)
4. L. Hou, S. Zhao, X. Xiong, K. Zheng, P. Chatzimisios, M.S. Hossain, W. Xiang, Internet of Things cloud: architecture and implementation. IEEE Commun. Mag. **54**(12), 32–39 (2016)
5. Q. Zhou, K. Zheng, L. Hou, J. Xing, R. Xu, Design and implementation of open LoRa for IoT. IEEE Access **7**, 100 649–100 657 (2019)
6. A. Mavromatis, C. Colman-Meixner, A.P. Silva, X. Vasilakos, R. Nejabati, D. Simeonidou, A software-defined IoT device management framework for edge and cloud computing. IEEE Internet Things J. **7**(3), 1718–1735 (2020)
7. F. Meneghello, M. Calore, D. Zucchetto, M. Polese, A. Zanella, IoT: internet of threats? A survey of practical security vulnerabilities in real IoT devices. IEEE Internet Things J. **6**(5), 8182–8201 (2019)
8. H. Dai, Z. Zheng, Y. Zhang, Blockchain for Internet of Things: a survey. IEEE Internet Things J. **6**(5), 8076–8094 (2019)
9. H.-T. Wu, C.-W. Tsai, An intelligent agriculture network security system based on private Blockchains. J. Commun. Netw. **21**(5), 503–508 (2019)

Chapter 2
Architecture of BIoT

2.1 Introduction

Directly using public blockchain in IoT, such as Bitcoin or Ethereum, is very expensive and inefficient. The IoT end-devices need to implement complex consensus algorithms, while all the data on chain are open to the public. In normal IoT system, the end-devices are only responsible for data sensing and uploading. To keep as long lifetime as possible, the end-devices need to be simple, and the operations must not consume too much resources and energy.

There are multiple participants involved in constructing the blockchain within the IoT system, including users, device owners, gateway owners, network providers, cloud computing service providers, database service providers, and application providers. Users generate the IoT data, which are collected by application providers to offer data services. The other participants collaborate to establish a channel for data transmission and exchange, ensuring efficient and secure communication within the IoT ecosystem. Therefore, to deploy blockchain in IoT system for data security and privacy, the blockchain needs to be specifically designed to fit the capabilities of IoT devices.

2.2 Related Works

In traditional IoT systems, data are stored in centralized servers, which may cause eavesdropping and single-point bottleneck. The blockchain can be used to build a distributed and trustful data storage method and let all related entities to participate and supervise for data security. Among most of the literature, blockchain for IoT becomes a widely studied scenario. Liang et al. [1] proposed a blockchain-based data transmission model for IoT. By using secure cryptologic algorithm and

© The Author(s), under exclusive license to Springer Nature Switzerland AG 2024
L. Hou et al., *Blockchain-Based Internet of Things*, SpringerBriefs in Computer Science, https://doi.org/10.1007/978-3-031-70303-4_2

decentralized consensus technique, the power data can be transmitted in a secure way. Wan et al. [2] deploy a private blockchain in IoT architecture in order to enhance the data security and privacy. The traditional storage layer is substituted by decentralized blockchain network. Moreover, Zhao et al. [3] propose a multichained IoT platform to provide secure and decentralized data transmission between IoT resources and end-users. Vehicular networks can also benefit a lot by blockchain. Kang et al. [4] deploy a Direct-Acyclic-Graph (DAG)-based blockchain network in vehicular networks with edge computing. DAG is considered as a new kind of data structure for blockchain, in order to enhance the efficiency. However, it may cause inconsistency after some time. Jiang et al. [5] propose a distributed Internet of vehicles architecture with blockchain. The transmission delay is analyzed considering the mobility of vehicles. Blockchain can also be deployed in healthcare scenarios. In order to avoid the health data being eavesdropped by third-party servers, a blockchain-based healthcare scheme is proposed where all end-devices, patients, and doctors participate in the blockchain network [6]. Blockchain can also improve the progress of machine learning for IoT. Qiu et al. [7] propose a collective Q-learning approach where the blockchain is used to share the training results between IoT nodes in order to govern the results in an efficient and secure way.

2.3 Multilayer and Multi-ledger Blockchain

In this monograph, a consortium blockchain with multiple ledgers is provided for the IoT applications. The consortium blockchain enables the join of all the IoT application participants and builds trust among them. To fully utilize the available resources of end-devices and base stations, the BIoT architecture is divided into three layers, i.e., **the cloud layer**, **the edge layer**, and **the end-device layer** as shown in Fig. 2.1. IoT end-devices are classified into **strong** and **weak** end-devices based on their computing capabilities. The blockchain is designed to be able to deploy on both IoT cloud servers, Multi-access Edge Computing (MEC) servers, and some of the strong end-devices. The layered approach distinguishes the performance characteristics of different IoT modules and meets the construction requirements of a multilayer blockchain network.

2.3.1 Multilayer Blockchain

The multilayer blockchain network consists of a global blockchain network in the cloud and several local blockchain networks. The global blockchain is maintained by servers in the cloud, which can be considered with unlimited computing and storage resources. While the local blockchains are built among strong end-devices and base stations, which are not capable of large-scale data storing and transaction

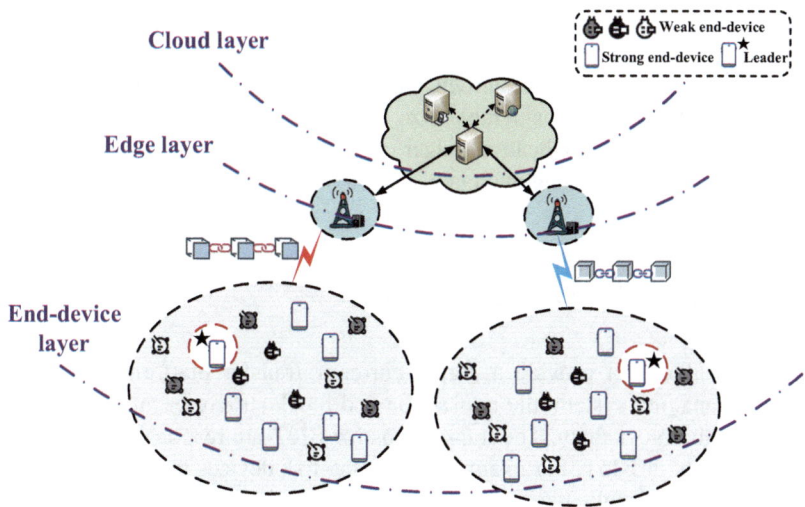

Fig. 2.1 Blockchain architecture for IoT applications

processing. Due to the different abilities, global blockchain keeps all the IoT data, and each local blockchain maintains only the IoT data generated in the local area. All the data are generated by weak end-devices, while the transactions and blocks are initialized and packaged by strong end-devices or base stations. The transactions are sent to both the local blockchain and global blockchain. Such a design is suitable for latency-sensitive and insensitive IoT applications. Latency-sensitive applications can be deployed on local blockchains as smart contracts. Because the local blockchains run at the edge, it is very close to the end-users. For those applications that are insensitive to latency or require a wide range of IoT data, they can be installed on global blockchain as smart contracts. Therefore, multilayer blockchain can meet the different requirements of IoT applications, while the different abilities of end-devices, edge servers, and cloud servers can also be utilized efficiently.

2.3.2   Ledger Structure

The ledger in BIoT stores all the IoT application-specific data, and these data are considered state of the BIoT applications. There are two major data types in IoT system, i.e., the network data that store the operation status, context, and device information and the application data that are application-specific and store application information for users. Therefore, this monograph provides two ledgers in consortium BIoT, i.e., **the network ledger** and **application ledger**. The network ledger is maintained by both the edge servers and cloud servers, while the application ledger is deployed only in cloud servers. This is because the network ledger stores network data whose size is limited. Besides, it requires frequent read and write for device registration, network access or handover, etc. Thus, deploying network ledger over the edge can reduce the latency of response. However, the IoT application data are of considerable size, which may exceed the ability of edge servers. Therefore, the application ledger can only be maintained in cloud servers. Both of these ledgers are deployed in global blockchain and local blockchains.

2.3.3 Transaction and Block Design

Since BIoT does not provide a digital currency transfer platform, but a secure IoT data managing system, the transaction and block structures are different from those in public blockchain. The transactions for BIoT are responsible for invoking smart contracts in BIoT. For example, when the IoT devices upload some sensing data to store, the application provider needs to initiate a new transaction, where the target smart contract, operation, and data are included in the payload field in the transaction. The transaction will be executed by BIoT nodes. After reaching a consensus, the data are written on chain and can be considered finality because they cannot be modified, revoked, or removed. The IoT data are in form of key–value pairs, for example, "Temperature: 20." All these data form a global state, where the state transition is triggered by all the transactions in the new block.

The block structure is similar to other blockchains. The block body consists of the list of transactions, while the block header is made up of several items such as the previous block header hash, the timestamp, the Merkle root hash of the transactions, and the Merkle root hash of the state of all the application data.

2.3.4 Consensus Algorithm

In public blockchain network, PoW and PoS are two of the most widely used consensus mechanism. PoW requires a lot of computing resources, while PoS is complex to implement. Neither of the consensus mechanisms is suitable for BIoT. The main reason is that these consensus mechanisms bring high workloads for IoT devices. The basic function of message exchanging may also be influenced. Therefore, how to reach consensus that both consider the fault tolerance and computing ability is one of the challenges in BIoT.

For permissioned blockchain, the distributed consensus mechanism RAFT is one of the suitable algorithms for consensus. RAFT is a leader-based consensus algorithm that functions with two stages, i.e., leader election and log replication. The leader in RAFT is elected randomly. A random timer is maintained by each node. When a node's timer times out, it becomes a candidate and initiates a leader vote request. The other nodes vote for the candidate. After receiving quorum votes, the candidate becomes leader. Leader status is maintained via heartbeat message. All client requests are handled by leader. The leader replicates the data to all the followers, and followers acknowledge the reception of data. When the leader gets quorum votes for one message, the message is considered as finalized. Therefore, consensus between distributed nodes is reached. RAFT consensus is lightweight compared with PoW, PoS, or PBFT, but it can only realize Crash Fault Tolerance (CFT).

To manage the load caused by deploying blockchain nodes on end-devices with varying capabilities, the network must employ an appropriate consensus mechanism. Traditional consensus algorithms, such as PoW [8] and PBFT [9, 10], are unsuitable for this architecture due to their high computational or communication requirements. To maintain the stable state of end-devices in the BIoT network, a lightweight blockchain consensus algorithm needs to be developed based on the performance characteristics of end-devices. Existing research on lightweight consensus algorithms can be divided into two categories. One category is to improve the common consensus algorithms [11, 12]. Based on PoW, Alhejazi et al. [13] proposed a novel algorithm adapted from the concept of weighted majority algorithm in ensemble learning, which enhanced the detection of malicious anomalies, thus improving the security level of IoT systems. Similarly, Huang et al. [14] proposed a credit-based PoW mechanism for IIoT scenario. To protect sensory data confidentiality, a data authority management method was designed in [14] based on directed acyclic graph-structured blockchains. Li et al. [15] proposed an improved PBFT consensus mechanism based on reward and punishment strategy. Meanwhile, a storage optimization scheme based on RS erasure code was proposed to reduce the cost of blockchain storage. Choi et al. [16] replaced the proof of work in PoW with the proof of trust accumulated by nodes working continuously in the blockchain network. Meanwhile, Zaman et al. [17] replaced the proof of work with the proportion of the resources paid by the nodes to the resources they have. Although the above improved methods can retain the mature algorithm flow of PoW, it is difficult to reduce the computational load caused by the consensus mechanism.

The other category is to design new consensus algorithms for specific application scenarios. Biswas et al. [18] simplified the consensus process into two steps, transaction verification and consensus formation. Peers were merged according to the number of nodes participating in the consensus to reduce the computing time required to reach consensus. For the consortium blockchain, Wang et al. [19] proposed a consensus fusion scheme based on trust levels to complete collaborative learning. The candidates with higher trust levels played a more important role in the consensus. Similarly, Kaci et al. [20] proposed a new reputation-based blockchain named PoolCoin based on a distributed trust model for mining pools. The trust model provided a machine learning module to assess the capabilities of mining machines, allowing pool managers to select trusted miners in their mining pools. In [21], each IoT sensor generated its unique tag and broadcasts it to the blockchain network. After a trusted node in the blockchain network authenticates the sensor, the sensor hashes the authenticated block through the lightweight hash function and adds the block to the blockchain. However, the RAFT algorithm has good adaptability to the blockchain network for IoT applications due to its algorithm characteristics [22, 23].

The RAFT algorithm divides the system running time into multiple terms, and the beginning of each term is leader selection. During leader selection, each node in the network switches among three node states according to leader selection results. The leader selection in the RAFT algorithm is triggered by the heartbeat mechanism. When any follower fails to receive the heartbeat packet sent by the leader within the fixed time, the follower converts to a candidate and triggers leader selection. In the selection, all the followers who fail to receive the heartbeat packet are selected as candidates. At the beginning of the selection, each candidate sends requests in parallel to obtain the votes of followers, and each follower votes for the candidate whose request it receives first. The candidate who obtains the votes of the majority nodes becomes the new leader. Additionally, the new leader sends heartbeat packets to other nodes in the network to establish leader authority and terminate the selection.

In a blockchain network with the RAFT algorithm as the consensus mechanism, the leader is responsible for packaging transactions to generate blocks, and blocks are sent to followers for endorsement. After receiving a new block, the followers endorse the block and return confirmation responses containing the endorsement results to the leader. The leader notifies nodes in the network to update the new block synchronously after receiving enough confirmation responses. The efficiency of this consensus algorithm is determined by the performance of the leader and the communication environment [24]. The block consensus method of the RAFT algorithm can avoid heavy computing burdens, while the unique leader ensures the consistency of data in the network [25]. However, the RAFT algorithm does not take into account the limited system resources in IoT applications since its selection parameters are fixed. Moreover, leader selection causes a large communication loss, and there is randomness in the leader selection process. To be implemented, the RAFT algorithm needs to be modified to fit the characteristics of various BIoT scenarios.

In all, RAFT is suitable for BIoT because in some scenarios of IoT, the participants are authenticated, and using RAFT can save a lot of resources while maintaining consensus on distributed ledgers.

2.3.5   Smart Contract

The core idea of blockchain is to construct a state replicated machine which is driven by the transactions inside blocks. The business logic of state transitions depends on the IoT applications, and the applications in BIoT are built in form of smart contract. Application providers develop IoT application smart contracts and install them to the blockchain, waiting to be invoked by clients or end-users. The smart contracts define the way to process transactions and change the state. To reach consensus, each smart contract needs to be deployed on all blockchain nodes. The smart contracts are invoked by transactions, since the smart contracts are deployed on all nodes, the state transitions are equivalent between these nodes, and thus, the consensus can be reached. As long as the transactions are initiated and the blocks are constructed, the states are transited definitely, and nothing can be done to call off such transitions. This is how the trust is established between all the participants, because they must be obliged to follow the rules defined in smart contracts. No one has the almighty power to modify the state discretionarily.

2.3.6   Example Procedure of the Provided BIoT System

For better understanding, an example of the blockchain procedure is provided below:

1. A new device needs to register to the network. The device needs to initialize a new transaction to call the registration smart contract with its information. The transaction, as well as other transactions from other devices, is received by the gateway or base station and stored in the transaction pool, waiting to be packed into block.
2. The gateway or base station retrieves a patch of transactions from the pool and sends to the leader according to RAFT consensus. After receiving the transactions, the leader begins to verify the validity of all transactions and executes each transaction to modify the global state.
3. By eliminating the invalid transactions, the leader packs a new block and sends an *appendEntries* message to all the followers.
4. All the followers execute all the transactions in the block and modify their global state to keep consistency with the leader. After successful execution, the followers reply the leader with an acknowledgement.
5. Suppose there are n nodes in total. When leader receives $\lceil n/2 \rceil$ acknowledge messages, the consensus is reached, and the block is considered in finality. The leader then sends messages to all followers to confirm the block. By receiving these messages, the followers chain the new block to their blockchain. Meanwhile, the leader responds the device that the registration is finished.

2.4   Advantages of Blockchain-Based IoT

The blockchain is suitable to provide data security and privacy for IoT. The BIoT system has the following advantages:

- **Access control:** To provide abundant applications for users, it is inevitable for applications to access a variety of data. The blockchain can offer access control smart contract for such cases that the applications can only read limited data within restricted time span under the permission of data owner. All the access records are stored on chain, and the records are immutable. Therefore, the data abuse risk is highly reduced.

- **Data assurance:** The data in BIoT is uploaded via transactions. The chained structure ensures that all the data stored are immutable, irrevocable, and non-repudiation. The transactions are organized by Merkle tree structure, and the hash root is stored in block header. Therefore, any modification on the data causes the changes of the entire successive block, which cannot reach consensus between all participants. Similarly, the transactions cannot be removed, which means the data owner cannot relocate or deny the data uploading transactions. The data can be declaimed as burned, which means data are permanently inaccessible from blockchain.

- **Trust:** The blockchain can construct trust between IoT system participants. This can be realized because all data and assets are kept on chain, and the operations are processed via smart contracts. The smart contracts define the rules and agreements and are also irrevocable and immutable. Therefore, once the smart contracts are deployed, all the participants must follow the rules, such as data sharing or assets transfer. With the enforceability of blockchain, IoT participants can cooperate with each other without prior trusts among them.

- **Edge computing:** The resources at the network edge need to be fully exploited to improve the quality of services. Additionally, the centralized paradigm needs to evolve into a decentralized one. Blockchain technology is well suited for IoT systems in this context.

In all, blockchain provides data security and privacy protection and constructs trusts among different IoT players. However, there are considerable challenges to deploy blockchain in IoT, which are mainly addressed in our monograph.

2.5 Summary

This chapter presents an architecture design for BIoT. To meet the requirements of IoT applications and fully utilize the resources of different entities in IoT, we introduce a multilayer blockchain with a multi-ledger design. The transactions and blocks are similar to those in a traditional blockchain, but the payload field in transactions is specific to IoT applications. Since BIoT can establish an authority management service, we have chosen RAFT as the consensus mechanism. The applications are implemented as smart contracts in BIoT, with application interfaces invoked by transactions from end-users. Our design enhances resource consumption efficiency while meeting the diverse requirements of various applications. However, there are still opportunities to reduce the latency of data being stored on the blockchain. The following chapters will provide more details on how to reduce latency during both the block construction and consensus phases.

References

1. W. Liang, M. Tang, J. Long, X. Peng, J. Xu, K. Li, A secure fabric blockchain-based data transmission technique for industrial Internet-of-things. IEEE Trans. Ind. Inf. **15**(6), 3582–3592 (2019)
2. J. Wan, J. Li, M. Imran, D. Li, Fazal-e-Amin, A blockchain-based solution for enhancing security and privacy in smart factory. IEEE Trans. Ind. Inf. **15**(6), 3652–3660 (2019)
3. S. Zhao, S. Li, Y. Yao, Blockchain enabled industrial internet of things technology. IEEE Trans. Comput. Social Syst. **6**(6), 1442–1453 (2019)
4. J. Kang, R. Yu, X. Huang, M. Wu, S. Maharjan, S. Xie, Y. Zhang, Blockchain for secure and efficient data sharing in vehicular edge computing and networks. IEEE Internet Things J. **6**(3), 4660–4670 (2019)
5. T. Jiang, H. Fang, H. Wang, Blockchain-based internet of vehicles: distributed network architecture and performance analysis. IEEE Internet Things J. **6**(3), 4640–4649 (2019)
6. J. Xu, K. Xue, S. Li, H. Tian, J. Hong, P. Hong, N. Yu, Healthchain: a blockchain-based privacy preserving scheme for large-scale health data. IEEE Internet Things J. **6**(5), 8770–8781 (2019)
7. C. Qiu, X. Wang, H. Yao, J. Du, F.R. Yu, S. Guo, Networking integrated cloud-edge-end in IoT: a blockchain-assisted collective Q-learning approach. IEEE Internet Things J. **8**(6), 12694–12704 (2021)
8. A. Gervais, G.O. Karame, K. Wüst, V. Glykantzis, H. Ritzdorf, S. Capkun, *On the Security and Performance of Proof of Work Blockchains* (Association for Computing Machinery, New York, 2016), pp. 3–16
9. H. Sukhwani, J.M. Martínez, X. Chang, K.S. Trivedi, A. Rindos, Performance modeling of PBFT consensus process for permissioned Blockchain network (Hyperledger Fabric), in *2017 IEEE 36th Symposium on Reliable Distributed Systems (SRDS)* (2017), pp. 253–255
10. Y. Meshcheryakov, A. Melman, O. Evsutin, V. Morozov, Y. Koucheryavy, On performance of PBFT Blockchain consensus algorithm for IoT-applications with constrained devices. IEEE Access **9**, 80 559–80 570 (2021)
11. C. Huang, Z. Wang, H. Chen, Q. Hu, Q. Zhang, W. Wang, X. Guan, RepChain: a reputation-based secure, fast, and high incentive Blockchain system via sharding. IEEE Internet Things J. **8**(6), 4291–4304 (2021)
12. S. Alrubei, E. Ball, J. Rigelsford, Securing IoT-Blockchain applications through honesty-based distributed proof of authority consensus algorithm, in *2021 International Conference on Cyber Situational Awareness, Data Analytics and Assessment (CyberSA)* (2021), pp. 1 7
13. M.M. Alhejazi, R.M.A. Mohammad, Enhancing the Blockchain voting process in IoT using a novel Blockchain weighted majority consensus algorithm (WMCA). Inf. Secur. J.: Global Perspect. **31**(2), 125–143 (2021)
14. J. Huang, L. Kong, G. Chen, M.-Y. Wu, X. Liu, P. Zeng, Towards secure industrial IoT: Blockchain system with credit-based consensus mechanism. IEEE Trans. Ind. Inf. **15**(6), 3680–3689 (2019)
15. C. Li, J. Zhang, X. Yang, L. Youlong, Lightweight Blockchain consensus mechanism and storage optimization for resource-constrained IoT devices. Inf. Process. Manage. **58**(4), 102602 (2021)
16. H.-S. Choi, G. M. Lee, W.-S. Rhee, Hierarchical trust chain framework for IoT services, in *2019 Eleventh International Conference on Ubiquitous and Future Networks (ICUFN)* (2019), pp. 710–712
17. M.U. Zaman, T. Shen, M. Min, Proof of sincerity: a new lightweight consensus approach for mobile Blockchains, in *2019 16th IEEE Annual Consumer Communications Networking Conference (CCNC)* (2019), pp. 1–4
18. S. Biswas, K. Sharif, F. Li, S. Maharjan, S.P. Mohanty, Y. Wang, PoBT: a lightweight consensus algorithm for scalable IoT business Blockchain. IEEE Internet Things J. **7**(3), 2343–2355 (2020)

19. K. Wang, C.-M. Chen, Z. Liang, M.M. Hassan, G.M.L. Sarné, L. Fotia, G. Fortino, A trusted consensus fusion scheme for decentralized collaborated learning in massive IoT domain. Inf. Fusion **72**, 100–109 (2021)
20. A. Kaci, A. Rachedi, Toward a machine learning and software defined network approaches to manage miners' reputation in Blockchain. J. Network Syst. Manage. **28**, 478–501 (2020)
21. S. Khan, W.-K. Lee, S.O. Hwang, AEchain: a lightweight Blockchain for IoT applications. IEEE Consum. Electron. Mag. **11**(2), 64–76 (2021)
22. D. Huang, X. Ma, S. Zhang, Performance analysis of the RAFT consensus algorithm for private Blockchains. IEEE Trans. Syst. Man Cybern.: Syst. **50**(1), 172–181 (2020)
23. D. Kim, I. Doh, K. Chae, Improved RAFT algorithm exploiting federated learning for private Blockchain performance enhancement, in *2021 International Conference on Information Networking (ICOIN)* (2021), pp. 828–832
24. D. Ongaro, J. Ousterhout, In search of an understandable consensus algorithm, in *2014 USENIX Annual Technical Conference (USENIX ATC 14)* (2014), pp. 305–319
25. D. Yu, W. Li, H. Xu, L. Zhang, Low reliable and low latency communications for mission critical distributed industrial Internet of things. IEEE Commun. Lett. **25**(1), 313–317 (2021)

Chapter 3
Transaction Migration Scheme for BIoT

3.1 Introduction

Data stored in public blockchain are the transactions information such as currency sender address, receiver address, and the request to call smart contracts. However, the data in IoT system consist of the environmental sensing data and control commands. How to design transaction and block structure to meet the requirements of IoT data storing and security is challenging.

With the aid of MEC, gateways and base stations can participate in blockchain networks and provide tamper-resistant data services at the edge [1]. However, the limited capabilities of these gateways and base stations pose significant challenges in handling the ever-increasing IoT data while meeting the low-latency requirements of time-critical applications [2]. When an anomaly is detected in a specific area, the IoT system must increase the measurement frequency of sensors to obtain high-resolution data for more accurate decision-making. Consequently, it takes more time to commit these data to the blockchain [3].

As the increasing of IoT data, the transactions may be stacked in the pool, especially when some emergent event happens at specific area. The sensing data in that area could raise a lot, causing the transactions be waiting in the pool. Therefore, the sensing data cannot be uploaded on chain timely, degrading the quality of services. In the following sections, we provide an intelligent transaction migration scheme to solve the problem. The scheme can migrate transactions between edge servers (or clusters), reaching a global minimum of data uploading time.

Therefore, this chapter provides an intelligent transaction migration scheme in which leaders can optimally migrate transactions to other clusters. In this scheme, the latency is defined as the sum of transaction migration latency, block generation latency, and block consensus latency. The optimization problem is formulated as a Markov Decision Process (MDP). To address the problem, we use a Deep Deterministic Policy Gradient (DDPG) based transaction migration scheme with action refinement, wherein a neural network generates the optimal policy [4, 5]. This

© The Author(s), under exclusive license to Springer Nature Switzerland AG 2024

L. Hou et al., *Blockchain-Based Internet of Things*, SpringerBriefs in Computer Science, https://doi.org/10.1007/978-3-031-70303-4_3

policy ensures that when sensing data in one area increase, incoming transactions can be optimally migrated to less congested areas. Simulations demonstrate that the provided scheme significantly reduces the time required to pack transactions into the blockchain.

3.2   Related Works

Many existing works have focused on minimizing the latency of blockchain-based IoT applications. Most of these studies adopt the PoW consensus mechanism and utilize MEC offloading to reduce workload [6]. However, PoW consensus wastes a significant amount of resources to resist malicious nodes. Rovira-Sugranes et al. [7] optimized latency by controlling sensor measurement rates in a DAG-based blockchain network. Nevertheless, DAG is unsuitable for distributed IoT applications because it does not guarantee consistency. Instead, RAFT, a promising consensus algorithm for distributed systems, is well suited for blockchain networks in IoT applications. It allows blocks to be finalized by a unique leader, avoiding heavy computational burdens and ensuring consistency. Although RAFT cannot prevent malicious nodes, an authentication center can manage the identification and authorization of IoT devices [8]. Some works address issues in blockchain networks using RAFT consensus. For example, Yu et al. [9] analyzed the latency and reliability of RAFT consensus in industrial IoT, while Xu et al. [10] studied the performance of RAFT under malicious jamming in wireless IoT environments.

3.3   System Model

Consider a RAFT-based BIoT system over MEC for IoT sensing application, as shown in Fig. 3.1. Assume there are K clusters in total, each of which maintains an independent ledger and can be indexed by $k = 1, 2, \ldots, K$. Assume base stations are elected as leaders because they are more reliable than gateways in this scenario. Therefore, the term *Leader k* represents the base station in the k-th cluster. Let V_k denote the number of gateways in cluster k. The data sensed by end-devices are sent to leaders as transactions and wait to be committed on ledgers. Let $\widehat{\tau}$ denote the epoch when the leader starts to generate new block. Assume new blocks are generated in fixed intervals and each gateway follows the leader's committing instructions and replicates the new block directly.

Leader k maintains a decision queue where transactions are waited to be processed at decision epoch τ. Assume end-devices sense the environment in fixed intervals. Then, the number of arriving transactions M_k is constant. At one τ, *Leader k* can take actions on N transactions at most. The actions can be either to move transactions to the local packing queue or to migrate transactions to other clusters. Transactions in the packing queue are encapsulated in a new block at next $\widehat{\tau}$. Let

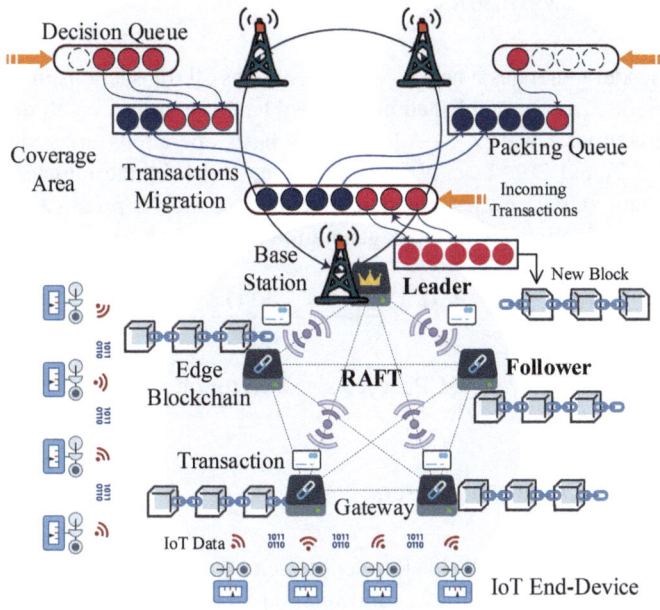

Fig. 3.1 Illustration of RAFT-based private blockchain with transaction migration for IoT applications

$\hat{\tau} = F\tau, F \in \mathbb{Z}^+$, which means block generation is triggered after F decision epochs. Assume the time interval between two decision epochs is fixed and is denoted as $\Delta\tau$. Let $\mathbf{a}_k(\tau) = \{a_{k,1}(\tau), a_{k,2}(\tau), \ldots, a_{k,N}(\tau)\}, \tau = 1, 2, \ldots$, denote the decision sequence of *Leader k* at τ. Let $a_{k,n} = 1, 2, \ldots, K, 1 \le n \le N$, where $a_{k,n} = k'$ means the transaction n needs to be migrated from *Leader k* to *Leader k'*. Typically, $a_{k,n} = k$ represents that the transaction is not migrated.

3.3.1 Transactions Migration Latency

The transactions migration latency can be computed as follows:

$$t_k^{\text{MIG}} = \frac{|I_k(\tau)|D}{G}, \tag{3.1}$$

where $I_k(\tau) = \{n | a_{k',n}(\tau) = k, k' \ne k\}$ and $|\cdot|$ represents the count of elements in set. D is the size of a transaction, and G is the transmission rate of the fiber link between two leaders. The transactions are directly put into the packing queue after being migrated. No more migration is allowed to avoid high latency.

3.3.2 Block Generation Latency

At $\hat{\tau}$, each leader generates a new block that contains all transactions in the packing queue. A Merkle tree is established by repeated hash operations on all transactions. For N transactions, $N + 2^{\mathcal{N}} - 1$ times of hash operations are required [11], where $\mathcal{N} \in \mathbb{Z}$ and $2^{\mathcal{N}-1} < N \leq 2^{\mathcal{N}}$. Let η_k stand for the number of central processing unit (CPU) cycles per unit time that *Leader k* possesses. Thus, the expected computing latency of block generation can be derived as follows:

$$t_k^{\text{COM}} = \frac{(N + 2^{\mathcal{N}} - 1)\xi}{\eta_k}, \tag{3.2}$$

where ξ stands for the average CPU cycles needed for one hash operation.

3.3.3 Block Consensus Latency

When a block is generated, the leader needs to send the *AppendEntries* message that contains the new block to all other followers to reach consensus. The transmission rate of the wireless link between *Leader k* and *Follower v* is defined as follows:

$$r_{k,v} = \frac{B}{V_k} \log(1 + \gamma_{k,v}), 1 \leq v \leq V_k, \tag{3.3}$$

where B is the total bandwidth, and it is equally allocated to V_k followers. Considering only the large-scale fading, the signal-to-noise ratio (SNR) of the wireless link from sender k to receiver v can be computed as follows:

$$\gamma_{k,v} = \frac{P d_{k,v}^{-\beta} \Psi}{B N_0}, \tag{3.4}$$

where P is the total transmission power that is equally divided for V_k followers. N_0 stands for the average power spectral density of white noise, and $d_{k,v}$ is the distance between *Leader k* and *Follower v*. β stands for the path loss ratio, and the $\Psi \sim \mathcal{LN}(0, \sigma^2)$ denotes the shadow fading which follows log-normal distribution.

To reach consensus on the new block, the leader has to receive confirmation messages from at least $\zeta_k + 1$ followers (including the leader itself), where $\zeta_k = \lfloor V_k/2 \rfloor$. Therefore, the consensus latency is determined by the transmission latency caused by the leader sending *AppendEntries* message to and receiving confirmation from the ζ_k-th follower, i.e.,

$$t_k^{\text{CON}} = \frac{C}{r_{k,\zeta_k}} + \frac{E}{r_{\zeta_k,k}}, \tag{3.5}$$

where C denotes the size of an *AppendEntries* message and E represents the size of the confirmation message.

3.3.4 Objective

Our objective is to minimize the long-term latency of all leaders by an optimal migration policy π^*. Giving consideration to fairness, the long-term latency is summed by the maximal latency of all clusters. Therefore, the objective problem is defined as follows:

$$\underset{\pi}{\text{minimize}} \sum_{\tau} \left\{ \max_{k} \left(t_k^{\text{COM}} + t_k^{\text{CON}} + t_k^{\text{MIG}} \right) \right\}. \tag{3.6}$$

3.4 DDPG-Based Transaction Migration Scheme

The optimization problem, Eq. (3.6), is formulated as an MDP model which consists of state set $\mathscr{S}$, action set $\mathscr{A}$, reward function $R : \mathscr{S} \times \mathscr{A} \to \mathbb{R}$, and transition probability $T : \mathscr{S} \times \mathscr{A} \times \mathscr{S} \to [0, 1]$. A reinforcement learning method is adopted to find the optimal policy π^* by a training procedure. Besides, an action refinement is applied to choose the best discrete action from the original output of a neural network.

3.4.1 MDP Model Formulation

3.4.1.1 State Set

System state S_τ is defined as $S_\tau = \{\mathbf{D}_\tau, \mathbf{P}_\tau, \Gamma_\tau, \iota\}$, where $\mathbf{D}_\tau = \{D_{1,\tau}, D_{2,\tau}, \dots, D_{K,\tau}\}$ represents the number of transactions in decision queue of each leader, and $\mathbf{P}_\tau = \{P_{1,\tau}, P_{2,\tau}, \dots, P_{K,\tau}\}$ stands for the status of packing queue, where $P_{k,\tau}$ includes transactions that come from local decision queue at previous epoch and that are migrated from other clusters, i.e., $P_{k,\tau} = N - |\{a_{k,n} | a_{k,n} \neq k, n \in [1, N]\}| + |I_k(\tau-1)|$. $\Gamma_\tau = \{\gamma_{1,1,\tau}, \dots, \gamma_{k,v_k,\tau}, \dots, \gamma_{K,v_K,\tau}\}$ denotes the SNRs of all wireless links. An indicator $\iota = \tau \mod F$ is used to indicate the number of decision epochs after a packing epoch. If $\iota = F$, a new block is generated. All possible states form the state set $\mathscr{S}$.

3.4.1.2 Action Set

An action $\mathbf{A}_{K \times N}(\tau)$ is given as a matrix, where each row stands for a decision sequence by the leader, i.e., $\mathbf{A}(\tau) = \{\mathbf{a}_k(\tau) | k \in [1, K]\}$. All possible actions form up an action set $\mathscr{A}$.

    Without extra explanations, the symbol τ is ignored for simplicity, i.e., $S_\tau := S$, $\mathbf{D}_\tau := \mathbf{D}, \mathbf{P}_\tau := \mathbf{P}, \Gamma_\tau := \Gamma$, and $\mathbf{A}_{K \times N}(\tau) := \mathbf{A}$.

3.4.1.3 State Transition Function

With an action $\mathbf{A}$ at a given state S, the following state can be determined via the state transition function. According to a different value of indicator ι in S, the transition function is defined as follows:

- **When** $\iota \neq F$, the transition function is given as follows:

$$S_{\tau+1} = f(S, \mathbf{A}) = \{D_1 + M_1 - N, \ldots, \tag{3.7}$$

$$D_K + M_K - N, P_{1,\tau+1}, \ldots, P_{K,\tau+1}, \iota + 1\}.$$

- **When** $\iota = F$, leaders need to collect all transactions in the packing queue to generate new blocks. Therefore, the transition function is given as follows:

$$S_{\tau+1} = f(S, \mathbf{A}) \tag{3.8}$$

$$= \{D_1 + M_1 - N, \ldots, D_K + M_K - N, 0, \ldots, 0, \iota = 0\}.$$

3.4.1.4 Reward Function

The instant reward is defined as the maximal latency of all leaders at τ. To keep consistency with the DDPG algorithm, the value of the reward function is set to be negative to make the larger reward stand for the lower latency. The reward function is given as follows:

$$R(\tau) = \begin{cases} 0, & \iota \neq F \\ -\max_{k \in K}\{t_k^{\text{COM}} + t_k^{\text{CON}} + t_k^{\text{MIG}}\}, & \iota = F. \end{cases} \tag{3.9}$$

When $\iota \neq F$, no block is generated. Therefore, the reward is set to be zero. Only at time $\hat{\tau}$ our model needs to calculate the latency.

3.4.2 Transaction Migration Scheme

Due to the large size of action space, the DDPG is adopted to make approximations on both action-value function Q and policy π [4]. The DDPG model includes an actor network and a critic network, each of which has a corresponding target network. Since the DDPG generates actions in continuous space, action refinement is adopted in our algorithm to choose the proper discrete action from the original continuous action, according to [12].

3.4.2.1 Actor and Critic Networks

Neural networks are used in DDPG to approximate both Q and π. Let $\hat{\pi}(S|\phi)$ denote actor network with parameter ϕ and $\hat{Q}(S, \mathbf{A}|\theta)$ denote the critic network with parameter θ. All layers in both networks are fully connected. The structures of target networks are equivalent to $\hat{\pi}$ and $\hat{Q}$, while the parameters are updated gradually via exponential moving average (EMA). Let ϕ' and θ' denote the parameters of target actor network and target critic network. Besides, let κ denote the rate of the EMA update on parameters. Then, the update function of target networks can be given as follows:

$$\phi' = \kappa\phi' + (1 - \kappa)\phi, \theta' = \kappa\theta' + (1 - \kappa)\theta. \tag{3.10}$$

For simplicity, define short notations for actor and critic networks as $\pi := \hat{\pi}(S|\phi)$, $Q := \hat{Q}(S, \mathbf{A}|\theta)$, and for target networks as $\pi' := \hat{\pi}(S'|\phi')$, $Q' := \hat{Q}(S', \mathbf{A}'|\theta')$. The loss $J(\phi)$ of the actor network is defined as the estimated value of Q that is computed by critic network, while the loss of the critic network is defined by the mean squared error of Q and the target value y which is given as follows:

$$y = R + \lambda Q', \tag{3.11}$$

where λ is the discount factor. Then, the loss of critic network can be computed as follows:

$$\mathscr{L}(\theta) = \mathbb{E}\{(Q - y)^2\}. \tag{3.12}$$

The main objective of DDPG is to find the optimal set of parameters θ^* to maximize Q which is calculated by the long-term sum of rewards, i.e., Eq. (3.6).

The detailed structure of both the actor network and the critic network is given in Table 3.1.

Table 3.1 Information of neural network structure

Network type	Layer type	Number of cells	Activation	Loss
Actor	Input Layer	$2K + 1 + \sum_k V_k$	–	Q
	Hidden Layer 1	8	ReLU	
	Hidden Layer 2	16	ReLU	
	Hidden Layer 3	8	ReLU	
	Output Layer	$K * N$	$K * \text{Sigmoid}$	
Critic	Input Layer	$2K + 1 + \sum_k V_k + K * N$	–	Mean
	Hidden Layer 1	8	ReLU	squared
	Hidden Layer 2	16	ReLU	error
	Hidden Layer 3	8	ReLU	
	Output Layer	1	–	

Algorithm 1: DDPG-based transaction migration training algorithm with action refinement

---

Input: Initialize parameters ϕ, $\theta \sim \mathscr{G}$, where $\mathscr{G}$ is a Gaussian distribution. Initialize a starting state $S = S_0 = \{0, 0, \ldots, 0\}$. Set experience deque $\mathscr{E} = \varnothing$, an index χ that indicate the epoch of training start and the total iteration times $\tau_{\max}$.

Output: π^*

while $\tau \neq \tau_{\max}$ **do**

    Sample a continuous action with random exploration, i.e., $\mathbf{A} = \pi(S|\phi) + g$, where $g \sim \mathscr{G}$;

    Refine actions according to Eq. (3.14);

    Each leader execute action $\mathbf{A}$ and observe the instant reward R ;

    Environment transit to state $S_{\tau+1} = f(S, \mathbf{A})$;

    Push the observation sequence $(S, \mathbf{A}, R, S_{\tau+1})$ into the experience deque $\mathscr{E}$;

    Set $\tau \leftarrow \tau + 1$ and $S \leftarrow S_{\tau+1}$;

    if $\tau > \chi$ **then**

        Randomly sample a mini-batch $e \sim \mathscr{E}$;

        Obtain target action batch by the target actor network via $\mathbf{A}' = \pi'(S)$, where $S \sim e$;

        Acquire the target action value batch by the target critic networks via $Q'(S, \mathbf{A}')$;

        Compute the value y according to Eq. (3.11);

        Calculate the gradient of the critic network as $\nabla \mathscr{L}$;

        Update parameters θ of the critic network towards the direction of gradient descent, i.e., $\theta \leftarrow \theta - \alpha_1 \nabla \mathscr{L}$, where α_1 is the learning rate of critic network;

        Get raw action batch by the actor network via $\mathbf{A} = \pi(S)$, where $S \sim e$;

        Calculate the gradient of the actor network as follows:

$$\nabla J = \frac{\partial Q}{\partial \phi} = \frac{\partial Q}{\partial \mathbf{A}} \frac{\partial \pi}{\partial \phi} \tag{3.13}$$

        Update parameters ϕ towards the direction of gradient ascent, i.e., $\phi \leftarrow \phi + \alpha_2 \nabla J$, where α_2 is the learning rate of the actor network;

        Update the parameters of the target networks according to Eq. (3.10);

    end

end

3.4.2.2 Action Refinement

The output of the actor network is the continuous features of actions, which cannot be directly executed by leaders. The most approaching discrete version generated by continuous action approximation may not lead to highest Q value [12]. Therefore, an action refinement, which can choose the discrete action to reach the maximal action-value function, is given as follows:

$$\mathbf{A}^* = \arg\max_{\hat{\mathbf{A}} \in \mathbb{A}} \hat{Q}(S, \hat{\mathbf{A}}), \tag{3.14}$$

where $\mathbb{A} = \{\tilde{\mathbf{A}} \big| \|\tilde{\mathbf{A}} - \pi\|_2 \leq \delta\}$. By using the proper value of δ, the balance between precision and efficiency can be reached.

3.4.2.3 Training Procedure

The training process of the transaction migration algorithm is given in Algorithm 1. After enough training, the actor network can generate optimal transaction migration policies for each leader of the cluster.

3.5 Performance Evaluation and Analysis

3.5.1 Simulation Setup

Simulations are conducted to model the scenario where the sensing frequency rises in some areas when an anomaly occurs and descends back to normal after the anomaly disappears, in an IoT application. There are four areas in total, and the anomaly occurs in *Area 2*. The number of incoming transactions in other areas stays to be 6, while M_2 varies from 4 to 9 and decays back to 4. Key parameters of our simulations are listed in Table 3.2. Each leader needs to receive $\zeta_k = 2$ responses from gateways to reach consensus by RAFT. Besides, the learning rates of both the actor network and the critic network are both set to be 10^{-5}.

3.5.2 Numerical Results and Analysis

Each simulation runs for 8×10^6 epochs. The actor loss $J(\phi)$ rises as the iteration time increases, while the critic loss $\mathscr{L}(\theta)$ is decreasing, as shown in Fig. 3.2. This is because actor loss $J(\phi)$ stands for the average action value of each batch. Thus, the higher $J(\phi)$ becomes, the higher rewards can be got. On the other hand, the critic loss $\mathscr{L}(\theta)$ represents the difference between the estimated action-value function and the target action-value function. Therefore, $\mathscr{L}(\theta)$ needs to decrease during the training process.

The provided scheme is compared to four baselines, i.e., random scheme, and three static schemes include non-migration, 30% migration, and 50% migration schemes. As the name implies, each baseline scheme defines how many transactions are migrated. The random scheme provides the average performance for all the possible schemes since each action of this scheme can be selected with equal

Table 3.2 Parameter configuration

Parameter	Value	Parameter	Value
K	4	V	4
B	20 MHz	N_0	150 dBm/Hz
P	24 dBm	β	3.4
ξ	300	η	1 GHz

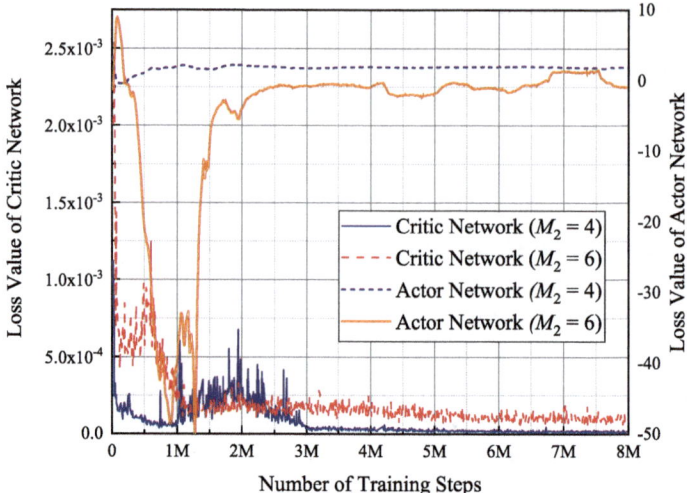

Fig. 3.2 Illustration of convergence of actor and critic networks under different M_2

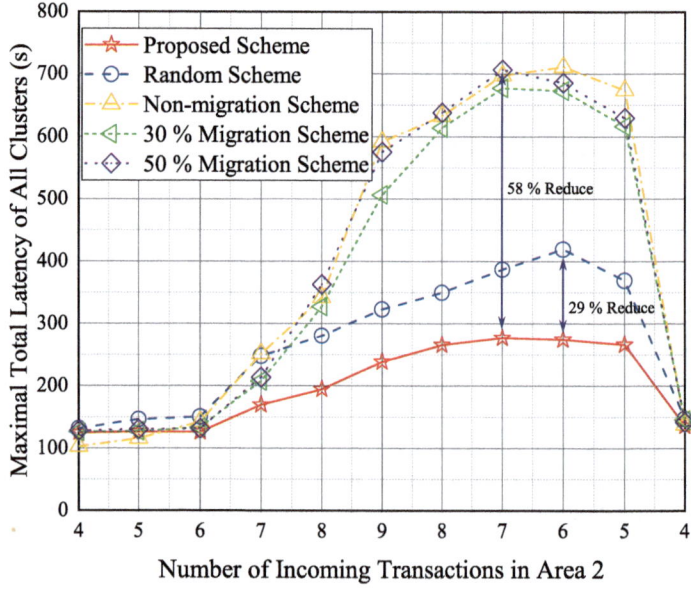

Fig. 3.3 Illustration of latency of all schemes under different M_2

probability after training enough times. As shown in Fig. 3.3, the three static baselines perform badly. It proves that static schemes cannot help improve the performance, even if they can migrate transactions from busy areas. Latency that is reached by the proposed scheme stays the lowest, because the proposed scheme can

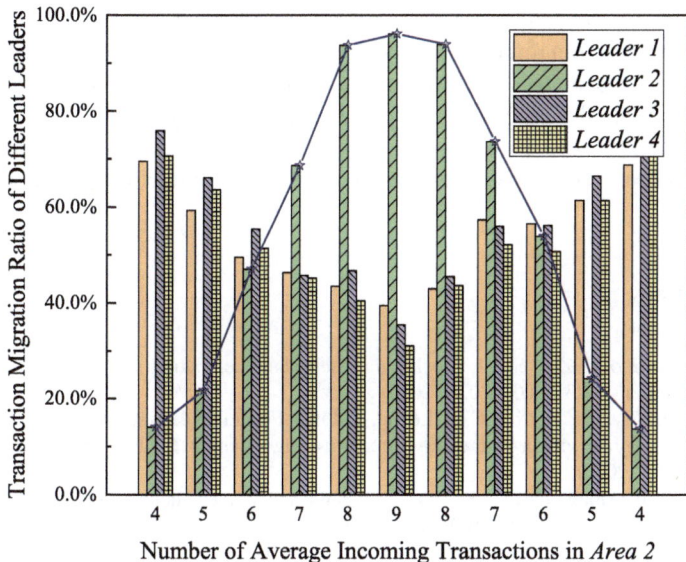

Fig. 3.4 Illustration of migration ratios of all leaders under different M_2

act intelligently with regard to the changing states. The proposed scheme balances the transactions among clusters optimally to reduce the latency by at most 58% and 29%, compared to the static schemes and random scheme.

As shown in Fig. 3.4, migration ratio of *Leader* 2 is the lowest at initial stage, because the workload is smaller than other three leaders. Therefore, the proposed scheme tends to migrate transactions from other leaders to reduce the maximal latency. When an anomaly occurs, the number of transactions rises in *Area* 2, causing *Leader* 2 to migrate transactions to other leaders as many as possible. Meanwhile, migration ratios of other leaders fall down to a low level. This is because our objective is to minimize the latency of the whole system rather than a particular cluster. Therefore, as the number of arriving transactions increases, the proposed algorithm tends to migrate transactions from the leader with high burden to the leaders that are idle. After the anomaly disappears, the migration ratios start to get back to normal.

3.6 Summary

In this chapter, we first give an architecture design of BIoT system, where the ledger, transaction, block, and consensus are discussed. Then, aiming at the latency requirement of IoT application, we provide a reinforcement learning-based scheme for RAFT-based consortium blockchain that intelligently migrates transactions between clusters. This scheme minimizes the latency of storing IoT data on the blockchain by balancing the workload among all clusters. We use the DDPG algorithm with action refinement to derive the optimal transaction migration policy, taking into account the dynamic status of the environment. Simulation results demonstrate that the proposed scheme can reduce latency by up to 58% when the system experiences an anomaly in a sensing area, compared to baseline methods. In the next chapter, a consensus mechanism for BIoT system will be discussed.

References

1. K. Gai, J. Guo, L. Zhu, S. Yu, Blockchain meets cloud computing: a survey. IEEE Commun. Surv. Tutorials **22**(3), 2009–2030 (2020)
2. X. Wang, C. Wang, X. Li, V.C.M. Leung, T. Taleb, Federated deep reinforcement learning for internet of things with decentralized cooperative edge caching. IEEE Internet Things J. **7**(10), 9441–9455 (2020)
3. P. Yuan, X. Xiong, L. Lei, K. Zheng, Design and implementation on Hyperledger-based emission trading system. IEEE Access **7**, 6109–6116 (2019)
4. T.P. Lillicrap, J.J. Hunt, A. Pritzel, N. Heess, T. Erez, Y. Tassa, D. Silver, D. Wierstra, Continuous control with deep reinforcement learning. arXiv e-prints:1509.02971 (2015), pp. 1–10
5. L. Lei, Y. Tan, K. Zheng, S. Liu, K. Zhang, X. Shen, Deep reinforcement learning for autonomous internet of things: Model, applications and challenges. IEEE Commun. Surv. Tutorials **22**(3), 1722–1760 (2020)
6. M. Liu, F.R. Yu, Y. Teng, V.C.M. Leung, M. Song, Computation offloading and content caching in wireless blockchain networks with mobile edge computing. IEEE Trans. Veh. Technol. **67**(11), 11008–11021 (2018)
7. A. Rovira-Sugranes, A. Razi, Optimizing the age of information for blockchain technology with applications to IoT sensors. IEEE Commun. Lett. **24**(1), 183–187 (2020)
8. C. Benzaïd, T. Taleb, AI for beyond 5G networks: A cyber-security defense or offense enabler?. IEEE Netw. **34**(6), 140–147 (2020)
9. D. Yu, W. Li, H. Xu, L. Zhang, Low reliable and low latency communications for mission critical distributed industrial Internet of things. IEEE Commun. Lett. **25**(1), 313–317 (2021)
10. H. Xu, L. Zhang, Y. Liu, B. Cao, RAFT based wireless blockchain networks in the presence of malicious jamming. IEEE Wireless Commun. Lett. **9**(6), 817–821 (2020)
11. L. Hou, K. Zheng, Z. Liu, X. Xu, T. Wu, Design and prototype implementation of a blockchain-enabled LoRa system with edge computing. IEEE Internet Things J. **8**(4), 2419–2430 (2021)
12. G. Dulac-Arnold, R. Evans, H. van Hasselt, P. Sunehag, T. Lillicrap, J. Hunt, T. Mann, T. Weber, T. Degris, B. Coppin, Deep reinforcement learning in large discrete action spaces. arXiv e-prints:1512.07679 (2015), pp. 1–11

Chapter 4
Lightweight Consensus Mechanism for BIoT

4.1 Introduction

As a decentralized system, BIoT requires a consensus mechanism to maintain data consistency [1]. The primary goal of the consensus mechanism is to achieve agreement on the generation of new blocks among all participants. To mitigate the impact of faulty and malicious nodes, the consensus mechanism must be crash fault-tolerant or even Byzantine fault-tolerant, without relying on third-party verification [2].

Lightweight consensus algorithms, such as RAFT [3], offer a viable solution to mitigate the computing load. As previously discussed, the RAFT algorithm, which features regular leader selection and leader-based block replication, is well suited for IoT systems [4, 5]. RAFT divides system runtime into terms, with nodes transitioning between three states: leader, follower, and candidate. Leader selection is triggered by the absence of heartbeat packets, and the candidate that secures a quorum of votes becomes the leader. The leader then sends heartbeat packets to other nodes to establish authority and end the selection process.

However, the RAFT algorithm does not consider the varying resources of IoT end-devices, as it relies on a purely random leader selection mechanism. This oversight can lead to resource wastage when applying RAFT in BIoT systems.

Therefore, this chapter introduces RAFT+, an improved one based on the RAFT algorithm with a new leader selection scheme, to enhance the performance of BIoT. First, RAFT+ collects computing resources of each end-device and samples the time-variant communication environment of the system as the parameters of leader selection. Then, a central node is chosen in the BIoT to collect selection parameters and select the leader. Considering the computing resources of different IoT entities, base stations or gateways are the central nodes in the local blockchain networks. Finally, all blockchain nodes participate in leader selection as candidates. The purpose of the above improvements is to make the optimal leader choice considering the current status of the system resources. The performance metric of

© The Author(s), under exclusive license to Springer Nature Switzerland AG 2024
L. Hou et al., *Blockchain-Based Internet of Things*, SpringerBriefs in Computer
Science, https://doi.org/10.1007/978-3-031-70303-4_4

leader selection is the block generation latency, which is defined as the sum of the block packaging latency and the block consensus latency. The optimization problem is formulated as an MDP model, which is solved by the Deep Q Network (DQN) algorithm.

4.2 Related Works

In the traditional IoT systems, end-devices and base stations are responsible for data reporting and forwarding, respectively, while computing services and databases are deployed on the central servers [6, 7]. The disadvantage of the centralized data processing structure is that system efficiency completely depends on the performance of the central servers [8, 9]. Meanwhile, there are security problems, such as privacy leakage [10].

Liu et al. [11] transferred part of the computing tasks to base stations, utilizing the computing resources of base stations to alleviate the load pressure on the central server and improve the efficiency of the system. Security challenges are generally addressed by combining IoT systems and blockchain in the existing research. In the blockchain network built for IoT systems, Guo et al. [12] utilized resources of all the IoT modules in the system to improve system efficiency. In specific IoT scenarios, the blockchain network needs to be designed according to the characteristics of scenarios to maintain the stable operation of the systems [13, 14]. In the IIoT scenario, network security is particularly important [15, 16]. Most IIoT infrastructures are based on a centralized architecture which is easier to manage but does not effectively support validation services between multiple parties. The blockchain-based IIoT architecture provides effective validation services and data storage schemes for resource-constraint IIoT infrastructures [17]. However, many prior works adopt common consensus algorithms, such as PoW, without considering the performance differences between the IoT modules. Due to the use of common consensus algorithms, a heavy workload is placed on end-devices. With the operation of the system, end-devices may stop working, thus affecting the integrity of the blockchain network and the efficiency of the system.

4.3 System Model

Consider the BIoT system that is previously illustrated in Fig. 2.1. It comprises a blockchain network deployed on the cloud and several local blockchain networks. The blockchain network on the cloud is composed of central servers. Each local blockchain network is constructed using a base station and strong end-devices within the base station's coverage area.

In this system, the local blockchain network employs the RAFT+ consensus mechanism, wherein the leader is responsible for block generation and initiating block consensus. When a block is generated within the local blockchain network, the leader notifies all nodes to synchronize and update with the new block. Subsequently, the base station forwards the newly generated block to the central server within the blockchain network on the cloud.

4.3.1 Scenario Description

Assume an IoT system with a single base station coverage scenario, where the number of strong end-devices and the number of weak end-devices are I and J, respectively. Strong end-devices are indexed by i, $i \in \{1, 2, \ldots, I\}$. A base station and strong end-devices form a blockchain network, of which the number of nodes is $I + 1$. The base station and strong end-devices are regarded as the central node and candidates of leader, respectively. The communication environment and computing resources of each strong end-device are considered during leader selection. After leader selection, the selected leader is represented by l, and the remaining nodes become followers. Weak end-devices collect and report data to the nearest strong end-device, while strong end-devices forward the collected data to leader l. Following the consensus algorithm, the data received by leader l is packaged in the form of transactions into blocks, and the new blocks are replicated to the blockchain ledger of each node in the blockchain network.

4.3.2 Communication Model

End-devices collect and report data at fixed intervals. Each end-device collects data for only one application type. The system time is divided into multiple epochs, and each epoch starts with the end of previous block consensus. Let n represent the number of data packets reported by each end-device in an epoch, which follows the Poisson distribution and is given as follows:

$$p(n) = \frac{\lambda^n}{n!} e^{-\lambda}, \tag{4.1}$$

where λ is the average number of data packets collected by the end-device in an epoch. D_m is the size of data packets, and the amount of data reported by the end-device in an epoch is $n D_m$.

During the data transmission process between the IoT modules, only large-scale fading including path loss is considered in the wireless channel model. The transmission rate between the IoT modules based on the Shannon formula is given as follows:

$$r_{l,i} = \frac{B}{I} \log_2 \left(1 + \gamma_{l,i}\right), \tag{4.2}$$

where B is the bandwidth used for block consensus in the system, and $\gamma_{l,i}$ is the Signal-to-Noise Radio (SNR) of data transmission between leader l and follower i, i.e.,

$$\gamma_{l,i} = \frac{P_{\text{CON}} d_{l,i}^{-\beta} \Psi}{B N_0}, \tag{4.3}$$

where P_{CON} is the total transmission power that is equally divided for the followers for block consensus, N_0 represents the average power spectral density of white noise, $d_{l,i}$ represents the distance between leader l and follower i, β stands for the path loss ratio, and $\Psi \sim \mathcal{LN}\left(0, \sigma^2\right)$ denotes the shadow fading which follows a log-normal distribution. An $I \times I$ matrix is used to represent the SNR of data transmission between the IoT modules in the blockchain network, which is given as follows:

$$\begin{pmatrix} 0 & \gamma_{1,2} \cdots \gamma_{1,I-1} & \gamma_{1,I} \\ \gamma_{2,1} & & \gamma_{2,I} \\ \gamma_{3,1} & & \gamma_{3,I} \\ \vdots & \ddots & \vdots \\ \gamma_{I-1,1} & & \gamma_{I-1,I} \\ \gamma_{I,1} & \gamma_{I,2} \cdots \gamma_{I,I-1} & 0 \end{pmatrix}. \tag{4.4}$$

4.3.3   Leader Selection Model

To enhance the overall communication efficiency in the process of block consensus, RAFT+ selects a central node in the blockchain network to collect the parameters of other nodes and selects the leader. If the leader does not receive a sufficient number of confirmation responses with endorsement results during the longest block consensus time $t_{\text{max}}^{\text{CON}}$, block consensus is considered failed, and the leader selection process will be triggered. There could be two reasons that lead to such a scenario, i.e.,

- **Hardware failure:** Hardware failure is defined as a hardware failure of the leader. Since if any follower has a hardware failure, the rest of the followers can still return confirmation responses, and the blocks can reach consensus. A hardware failure of the leader causes the leader to fail to receive confirmation responses, resulting in a failed block consensus. The probability of such a fault is represented by p_{hard}.
- **Communication failure:** Data transmission rate $r_{l,i}$ between leader l and follower i is determined by the SNR, and SNR $\gamma_{l,i}$ between leader l and follower i changes with the time-variant communication environment of the system. When data transmission rate $r_{l,i}$ falls below the minimum rate requirement,

the communication between nodes is interrupted. Data transmission rate $r_{l,i}$ falls below the minimum requirement, which means SNR $\gamma_{l,i}$ falls below the minimum SNR requirement. The communication interruption probability $p_c\,(l,i)$ between leader l and follower i is expressed as follows:

$$p_c\,(l,i) = \mathrm{Pr}(\gamma_{l,i} < \gamma_0) = \int_0^{\gamma_0} f(x)dx, \qquad (4.5)$$

where γ_0 represents the minimum SNR requirement for data transmission between nodes, and $f(x)$ is the probability density function of SNR. If more than $\lceil I/2 \rceil$ followers meet the communication interruption with the leader, communication failure occurs in the blockchain network. Let B_{tr} represent the set of all followers at timestamp t and B_{er} represent the set of followers which meet the communication interruption during data transmission at timestamp t, $B_{er} \subseteq B_{tr}$. When the number of elements in set B_{er} is more than $\lceil I/2 \rceil$, the probability p_{com} of communication failure due to communication interruption is expressed as follows:

$$p_{com} = \prod_{j \in B_{er}} p_c\,(l,j) \cdot \prod_{m \notin B_{er} \cap m \in B_{tr}} [1 - p_c\,(l,m)]. \qquad (4.6)$$

The probability of triggering the leader selection process by various faults is expressed as follows:

$$p_e = p_{hard} + p_{com} - p_{hard} \cdot p_{com}. \qquad (4.7)$$

After completing leader selection, the central node sends the selection result to nodes in the blockchain network. Each node updates its node status and resends the reported but unpackaged data to the new leader. The new leader completes block packaging and block consensus tasks.

4.3.4 Block Consensus Model

The basic block consensus flow of RAFT+ is shown in Fig. 4.1. The data reported by weak end-devices and forwarded by strong end-devices are stored in the cache pool of the leader in the form of transactions. The leader packages all transactions in the cache pool at the packaging interval to generate blocks. The number of transactions contained in the block is determined by the amount of data reported by each end-device. After the block is generated, the leader sends it to the followers in the blockchain network for endorsement. Each follower returns a confirmation response containing the endorsement result. After the leader receives $\lceil I/2 \rceil$ confirmation responses, the generation of the block reaches a consensus within the blockchain network, and the leader notifies the followers to synchronously update the block.

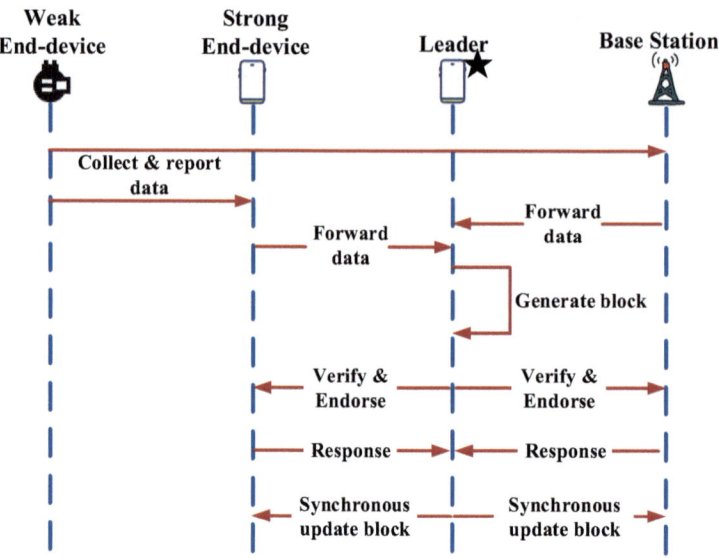

Fig. 4.1 The basic block consensus flow

The block generation process is mainly divided into two steps, i.e., block packaging and block consensus.

During a block consensus process, if the leader fails to receive enough confirmation responses, block consensus will be invalid, and leader selection will be triggered. There are two main reasons for the above situation, one is the leader's hardware failure, and the other is the poor communication environment of the system. When the new leader is selected through the leader selection scheme, followers will report the data that fail to generate the block to the new leader. The new leader will package the data to generate a new block and start block consensus. Leader selection in RAFT+ is dominated by the central node, which collects the parameters of each node in the blockchain network and selects the new leader. The new leader collects and packages data in the form of transactions to generate a new block. After the new block is generated, block consensus is carried out in the network. The specific process is shown in Fig. 4.2.

In the provided blockchain architecture, a block is composed of a block header and a block body. The block header mainly stores metadata for identifying blocks, and the Merkle root is calculated based on the transactions stored in the block. Specifically, the Merkle root is generated by the Merkle tree, and each leaf node in the Merkle tree represents a transaction stored in the block. For block k containing n_k transactions, the number of hash values n_s contained in the Merkle root is given as follows:

$$n_s = 2n_k - 1. \tag{4.8}$$

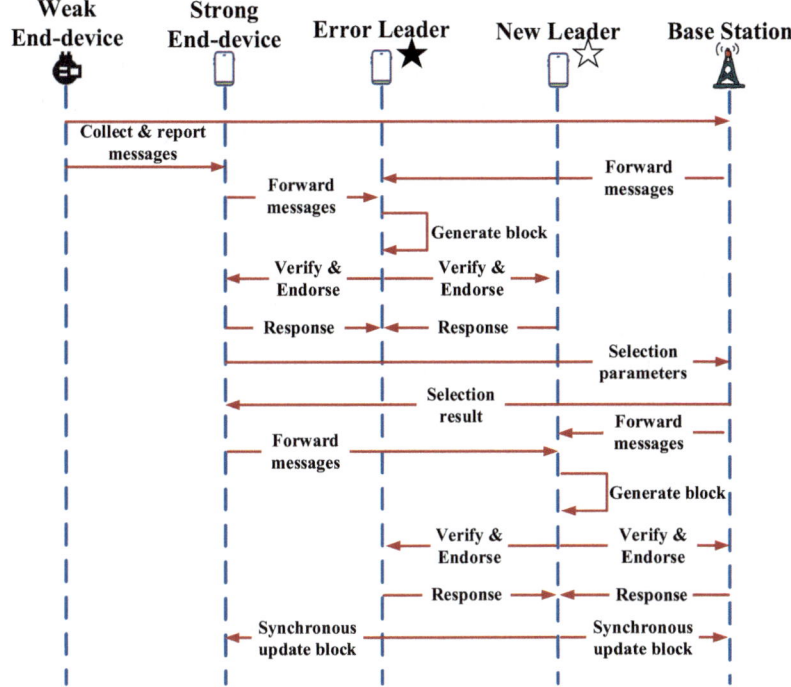

Fig. 4.2 The block consensus flow when leader failures occur

The block body stores specific data in the form of transactions. The amount of data reported by strong end-device i received in block k is represented by $n_{k,i}$, and the number of transactions stored in block k is given as follows:

$$n_k = \sum_{i=1}^{I} n_{k,i}. \tag{4.9}$$

Let η_l indicate the number of Central Processing Unit (CPU) cycles available for leader l in a unit time and ξ_h represent the number of CPU cycles required to calculate a hash value from the hash function for a transaction. For block k with n_k transactions, $n_k - 1$ times of hash operations are required to generate the Merkle root. The number of CPU cycles required for a hash operation to generate the Merkle root is ξ_m. Combine the above variables; the block packaging latency is given as follows:

$$t_{l,k}^{\text{PACK}} = \frac{n_k \xi_h + (n_k - 1)\xi_m}{\eta_l}. \tag{4.10}$$

In the RAFT algorithm, the leader sends heartbeat packets to followers in the blockchain network to maintain the state of each node. In RAFT+, the new blocks are regularly packaged by leader l and reported to followers also serve to maintain the state of each node. The data transmission process of block consensus includes two parts, which are (i) the leader sends the new block to the followers and (ii) the followers return confirmation responses containing the endorsement results. The block consensus latency between leader l and follower i can be expressed as follows:

$$t_{i,k}^{CON} = \frac{n_k D_t + n_s D_h}{r_{l,i}} + \frac{D_f}{r_{i,l}} = \frac{n_k D_t + n_s D_h + D_f}{r_{l,i}}, \tag{4.11}$$

where D_t represents the data size of a transaction, D_h represents the data size of a hash value, and D_f represents the data size of a confirmation response. Assuming that the confirmation response returned by follower i is the $\lceil I/2 \rceil$th confirmation response received by the leader, the generation latency for block k is given as follows:

$$t_{l,k}^{GEN} = t_{l,k}^{PACK} + t_{i,k}^{CON}. \tag{4.12}$$

4.3.5 Objective

The optimization goal of the system is to obtain the optimal leader selection strategy π^* for RAFT+. In the strategy, selection parameters are comprehensively analyzed to choose a leader that can minimize the average block generation latency. The specific optimization problem can be defined as follows:

$$\min \mathbb{E}_\tau \left[\frac{1}{K} \sum_{k=1}^{K} \left\{ (1 - p_e) t_{l,k}^{GEN} + p_e \left(t_{\max CON} + t_{z,k}^{GEN} \right) \right\} \right],$$

$$\text{s.t. C1} : l \in \{1, 2, \dots, I\}, z \in \{1, 2, \dots, I\},$$

$$\text{C2} : l \neq z, 0 \leq p_e \leq 1, \tag{4.13}$$

where z represents the new leader selected by the central node. The confirmation response returned by follower i is the $\lceil I/2 \rceil$th confirmation response that the leader has received. $t_{l,k}^{GEN}$ represents the block generation latency of block k generated by leader l, and $t_{z,k}^{GEN}$ represents the block generation latency of block k generated by new leader z.

4.4 DQN-Based RAFT+ Algorithm

The optimization problem in (4.13) can be formulated as an MDP model which consists of a state set $\mathscr{S}$, an action set $\mathscr{A}$, and a reward function $R : \mathscr{S} \times \mathscr{A} \to \mathbb{R}$. Let τ denote the epoch when the leader starts to generate a new block. Without extra explanations, the symbol τ is omitted for simplicity, i.e., $S_\tau := S$, $A_\tau := A$. The DQN algorithm is adopted to find the optimal policy π^* by a training procedure.

4.4.1 MDP Model Formulation

4.4.1.1 State Set

System state S is defined as follows:

$$S = (M, L, P, N, E), \tag{4.14}$$

where M represents the number of data received by the leader. Let M_i represent the number of data received by each strong end-device at epoch τ. M can be specifically expressed as follows:

$$M = \sum_{i \in \{1,2,\dots I\}} M_i. \tag{4.15}$$

L represents the serial number of the leader in the blockchain network, i.e., $L = l$. P represents the probability of hardware failure occurred at the leader, i.e., $P = p_{\text{hard}}$. N represents the SNR values between the leader and the followers in the blockchain network during data transmission, which can be obtained from (4.4). E represents the node state of the leader at epoch τ: $E = 0$ means that the leader does not need to be replaced, and $E = 1$ means that the leader needs to be replaced.

4.4.1.2 Action Set

In this model, action A taken at epoch τ is defined as the serial number of the selected leader, which can be expressed as follows:

$$A = l, \tag{4.16}$$

where $l \in \{1, 2, \dots, I\}$. The block packaging interval is represented by t_p, and the model generates actions at intervals of t_p. If more than $\lceil I/2 \rceil$ followers return confirmation response in the process of block consensus, the leader is functioning normally, and action A_τ remains the same as action $A_{\tau-1}$ at the last epoch. All possible actions form action set $\mathscr{A}$.

4.4.1.3 State Transition Function

With an action A at a given state S, the following state S' can be determined via the state transition function. The system state $N_{\tau+1}$ at the next epoch can be expressed as follows:

$$N_{\tau+1} = \left\{ \gamma_{l,i}^{\tau+1} | i \in B_{tr} \right\}. \tag{4.17}$$

According to system states L_τ, $N_{\tau+1}$ and $P_{\tau+1}$, the failure of leader l is examined; thus system state $E_{\tau+1}$ at the next epoch is determined. The specific system state $E_{\tau+1}$ can be expressed as follows:

$$E_{\tau+1} = \begin{cases} 0, \text{ normal} \\ 1, \text{ error.} \end{cases} \tag{4.18}$$

Assuming that the selected leader at epoch τ is A_τ, the state transition function of system state L can be obtained according to $E_{\tau+1}$, which can be expressed as follows:

$$L_{\tau+1} = \begin{cases} L_\tau, E_{\tau+1} = 0 \\ A_\tau, E_{\tau+1} = 1. \end{cases} \tag{4.19}$$

4.4.1.4 Reward Function

The instant reward is defined as the block generation latency at epoch τ, which consists of the block packaging latency and the block consensus latency. The reward function is given as follows:

$$R(\tau) = \begin{cases} t_{l,k}^{\text{PACK}} + t_{i,k}^{\text{CON}}, & E_\tau = 0 \\ t_{\text{max}}^{\text{CON}} + t_{z,k}^{\text{PACK}} + t_{i,k}^{\text{CON}}, & E_\tau = 1. \end{cases} \tag{4.20}$$

4.4.2 Leader Selection Scheme

In the MDP model established in this chapter, the system states are discrete, and the number of the states is limited. Meanwhile, the size of the action set is determined by the number of strong end-devices. So this chapter adopts the DQN algorithm to learn function Q and strategy π^*. The DQN algorithm combines reinforcement learning with deep neural networks and utilizes an experience replay strategy to eliminate the correlation between input sequences. The DQN algorithm includes two neural networks, i.e., the estimation network and the target network. The estimation network is trained in real time according to a properly defined loss function. The target network uses the weights of the estimation network at intervals to update its network parameters, and the target network is used to calculate the objective function value. The two neural networks are used independently to eliminate the correlation between the estimated Q value and the target Q value. As the central node, the base station collects selection parameters stored in each follower and selects the leader according to strategy π^*, which can be expressed as follows:

$$\pi^*(S, A) = \arg\max_{A \in \mathscr{A}} Q(S, A). \tag{4.21}$$

The DQN algorithm uses the following formulas to update the parametric loss function $\mathscr{L}(\theta)$ during iterations:

$$\mathscr{L}(\theta) = \mathbb{E}[(Q(S, A, \theta) - Q_{\text{Tar}})^2], \tag{4.22}$$

$$Q_{\text{Tar}} = R(S, A) + \gamma \max_{A' \in \mathscr{A}} Q(S', A', \hat{\theta}), \tag{4.23}$$

where θ represents the parameters of the estimation network, $\hat{\theta}$ represents the parameters of the target network, and γ is a discount coefficient, $0 \leq \gamma \leq 1$. S' and A' are the system state and the action at the next epoch. RAFT+ is based on the DQN algorithm, which is shown in Algorithm 2. After sufficient training, neural networks can generate the optimal leader selection strategy.

4.5   Performance Evaluation and Analysis

4.5.1   Simulation Setup

To evaluate the performance of RAFT+, we developed a simulation platform based on Python. In the simulations, IoT end-devices collect and report perception data according to preset intervals. Meanwhile, the location of IoT end-devices is fixed, and all IoT end-devices are directly connected to the power supply. The amount of data generated in the area within a fixed time is constant, and this chapter assumes that the energy consumption generated by each IoT end-device performing the block packaging task is the same. In the blockchain network with RAFT+ as the consensus mechanism, blockchain nodes do not need to participate in mining, thus reducing the number of computing tasks for each node and greatly reducing the system energy consumption. Due to the limited storage resources of IoT end-devices, this chapter adopts the method of building a multilayer blockchain network to control the scale of the local blockchain network to minimize the amount of data that strong end-devices need to store. The number of strong end-devices is determined by the specific simulation settings. Meanwhile, the number of weak end-devices is consistent with the number of strong end-devices. End-devices are randomly distributed in the simulation area with a radius of 100 meters to collect sensory data at fixed intervals. Weak end-devices collect and report the data to the closest strong end-devices, and strong end-devices forward the data to the leader

---

Algorithm 2: DQN-based RAFT+ algorithm

Input: I

nitialize p_{hard}, Ψ and the number of end-devices. Initialize neural network parameters ε, γ, θ, $\hat{\theta}$, and learning rate α. Set start time of iteration and the total iteration time τ_{max}. Set update timestep t_e of the target network;

Output: s

trategy π^*;

Initialize a starting action $A = A_0 = 0$ and a starting state $S = S_0 = (M_0, L_0, P_0, N_0, E_0)$, building neural Networks;

while $\tau \neq \tau_{\text{max}}$ **do**

    Update the real-time SNRs of data transmission between end-devices;

    Update the state N according to the SNRs;

    According to the system state, determine whether the system has a hardware failure or communication failure;

    if *Hardware failure or communication failure* **then**

        Update state E, generate random number e, start leader selection;

        if $e \geq \varepsilon$ **then**

            Generate Q values corresponding to all actions according to the estimation network;

            A new leader is selected according to the (4.21) from Q values;

        end

        else

            Keep the same action as the previous epoch;

            Randomly select new a leader as action A_τ;

        end

        Update state $L_\tau = A_\tau$;

    end

    else

        Update state $L_\tau = L_{\tau-1}$;

    end

    Each end-device execute action A_τ and leader observe R and $S_{\tau+1}$;

    Push the observation sequence $(S_\tau, A_\tau, R, S_{\tau+1})$ into the experience replay pool;

    Sample S, S', A and R from the experience replay pool;

    The estimation network and the target network respectively predict Q value according to S and S';

    Update the corresponding Q value in the estimation network based on R and predicted Q values according to (4.22);

    Update loss function $\mathscr{L}(\theta)$ and parameter θ of the estimation network according to (4.23);

    if τ mod $t_e = 0$ **then**

        Update the target network parameter $\hat{\theta}$ according to the estimation network parameter θ;

    end

    Set $\tau + 1 \to \tau$ and $S_{\tau+1} \to S_\tau$;

end

---

in the blockchain network. At each packaging time, the leader packages the data received since the last packaging time to generate a block and sends the block to the blockchain network. RAFT+ serves as the consensus mechanism of the blockchain network. The total available bandwidth of the system is 10 MHz. The system bandwidth is used for data reporting by weak end-devices, data forwarding by strong end-devices, and data transmission between the leader and the followers when block consensus occurs. The SNR of data transmission between the leader and each follower is determined by system state N at that epoch. Only the large-scale fading propagation is considered in our simulations, and path loss is considered in the wireless channels. The main simulation parameters of the IoT system are shown in Table 4.1.

Each training round of the system completes 10^6 epochs (steps). In order to achieve optimal simulation results within a long simulation epoch, the learning rate of neural networks in DQN is set as 10^{-3}. The specific parameters of neural networks are listed in Table 4.2.

Table 4.1 System simulation parameters and values

Parameter	Value
Area radius	1 km
Number of strong end-devices	$3, 4, \ldots, 12$
Number of weak end-devices	$3, 4, \ldots, 12$
CPU cycles of end-devices for hash operations	20, 40, 60 MHz
Number of messages reported by end-devices	$2, 3, \ldots, 11$
Hardware failure probability	5%
System bandwidth	10 MHz
Bandwidth for each weak end-device	200 kHz
Bandwidth for each strong end-device	400 kHz
SNR threshold	-20 dB

Table 4.2 Neural network parameters and values

Parameter	Value	Parameter description
Layers of neural network	3	Number of layers of neural network
Neurons per layer	128	Number of neurons in each layer of neural networks
Training steps	10^6	Number of training steps per simulation
Learning rate	10^{-3}	Control the learning progress during iterations
Reward_decay	0.9	Discount rate for long-term reward
ε_greedy	0.8	Probability of choosing the action with the highest Q value
Replace_target_iter	300	Interval of replace the target network parameters with the estimation network parameters
Memory size	500	Cache pool size
Batch size	32	Size of the sample from cache pool

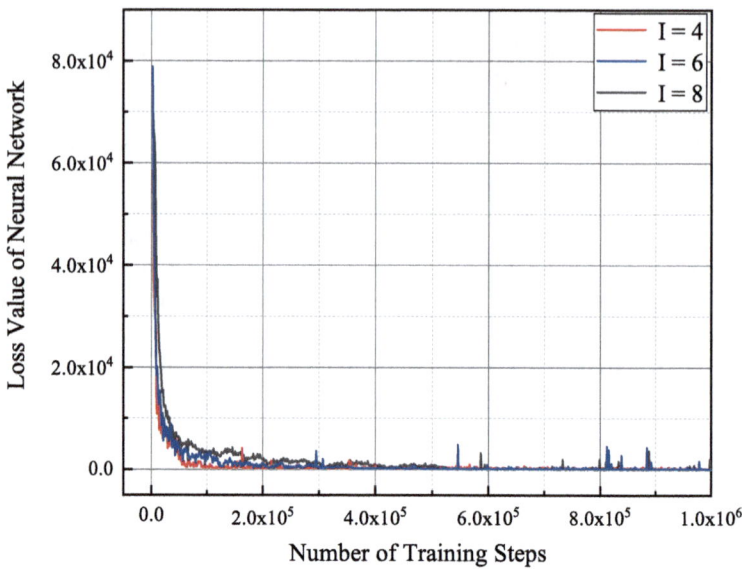

Fig. 4.3 The convergence of estimation network with different numbers of end-devices

4.5.2   *Numerical Results and Analysis*

As the number of iterations increases, the leader selection strategy obtained by neural networks is gradually approaching the optimal strategy. Figure 4.3 shows the values of loss under different numbers of end-devices, where I represents the number of strong end-devices in the simulations. The smaller the loss value, the closer the predicted value to the target value. It means that more training steps are required for the convergence of the loss function with the increase of end-devices in the system. Since there are two neural networks in the DQN algorithm for independent training and regular updating of neural network parameters, as well as the SNR values between end-devices are randomly distributed, the loss values fluctuate to a certain extent after the convergence of the loss function. Simulation results show that the loss values of the estimation network are from 10^4 to 10^2, and the average number of iterations is about 10^5.

In order to show the impact of different consensus mechanisms of the blockchain network on system performance, RAFT+ is compared with two baseline schemes, i.e.,

- **Random selection scheme:** Before each block packaging moment, the leader is selected randomly from all the nodes in the blockchain network to complete block generation and lead block consensus.
- **RAFT:** The original RAFT algorithm determines whether the leader fails at each block packaging time. In particular, when the original RAFT algorithm is used as the consensus mechanism of the blockchain network, all followers are selected as

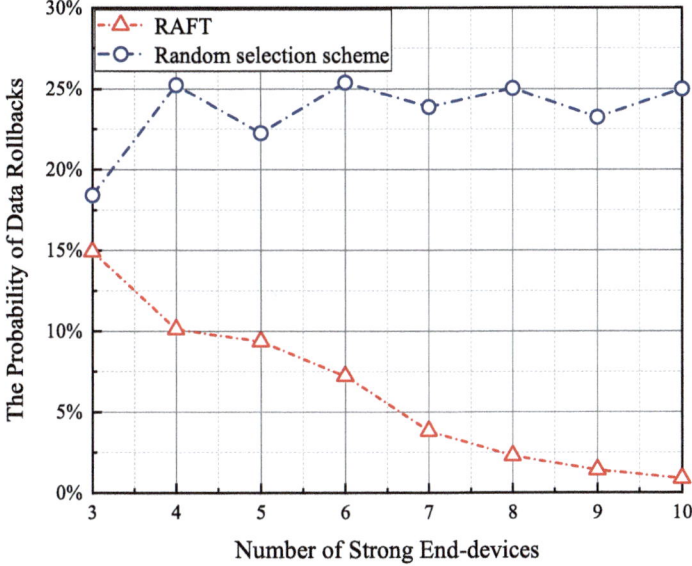

Fig. 4.4 The probability of data rollback under different schemes

candidates to make an intuitive comparison. If the original leader fails, the leader is reselected according to certain parameters, such as term and index.

For the sake of comparison, data rollback is defined as the case that a follower's data are overwritten due to data inconsistency between the leader and the followers. In the blockchain network, data rollbacks may occur due to hardware failure or communication failure. During the block consensus process, if the leader notices that the blocks stored in a follower are inconsistent with the leader, the inconsistent blocks in the follower will be deleted. Meanwhile, the follower's ledger will be updated to be consistent with the leader's ledger. Figure 4.4 shows the probability of data rollbacks under different consensus mechanisms. Due to the unique leader selection scheme, the RAFT algorithm can significantly reduce the probability of data rollbacks with an increase in the number of end-devices. The random selection scheme cannot limit data rollbacks, resulting in data loss. RAFT+ inherits the characteristics of the RAFT algorithm, which can minimize the possibility of data rollbacks.

Figure 4.5 shows the impact of different consensus mechanisms of the blockchain network on the average block generation latency. In the simulations, the amount of data reported by each end-device is gradually increased. In Fig. 4.5, I represents the number of strong end-devices in the simulations. It is shown that the average block generation latency rises gradually with an increase in the amount of data reported by end-devices. Compared with the RAFT algorithm, the provided RAFT+ algorithm can reduce the average block generation latency by 10% from 21%. Meanwhile, with the increased amount of data reported by the end-devices, the impact of

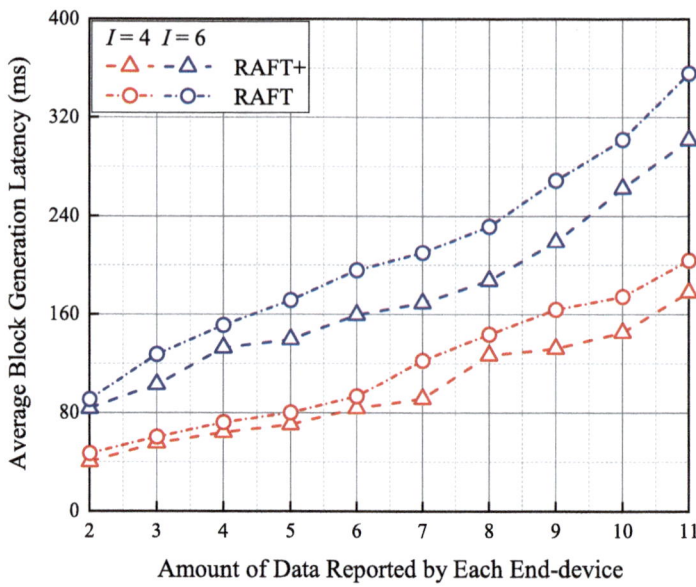

Fig. 4.5 The average block generation latency with different amount of data reported by each end-device

RAFT+ on reducing the average block generation latency becomes gradually obvious. The above results indicate that RAFT+ as the consensus mechanism can effectively improve the operating efficiency of the system and the load capacity of the blockchain network.

Figure 4.6 shows the impact of different consensus mechanisms of the blockchain network on the average block generation latency under different numbers of IoT end-devices. In the simulations, the number of strong end-devices is increased from 3 to 12, and each end-device reports two data units in each epoch. In the simulations, the computing resources of each strong end-device are independently selected in a certain range, which directly results in the fluctuation of the average block generation latency. Though the amount of data rises with the increase of end-devices, the computing resources of strong end-device still affect the block packaging latency and lead to the fluctuation of the average block generation latency. As the number of strong end-devices is increased, RAFT+ can maintain a low average block generation latency. When IoT end-devices have stronger computing resources, the system efficiency is further improved. RAFT+ provides an approximately 15% performance improvement over the RAFT algorithm. The performance improvement is achieved because the computing resources of strong end-devices and the time-variant communication environment of the system are not taken into account in the RAFT algorithm. However, the leader selection scheme in RAFT+ comprehensively considers the IoT system resources and obtains the optimal strategy through neural network training. It can be seen that the minimum average block generation latency can be obtained by using RAFT+ according

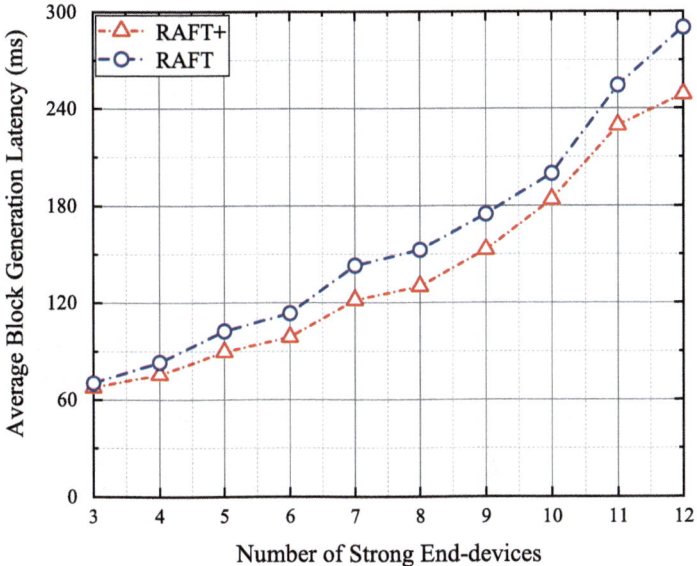

Fig. 4.6 The average block generation latency with different numbers of strong end-devices

to Figs. 4.5 and 4.6. The simulation results indicate that RAFT+ is scalable. Considering the composition of the local blockchain network and the limited computing resources of IoT end-devices, the number of nodes in the blockchain network is limited. Meanwhile, with the increase of the number of IoT end-devices, the simulation time increases dramatically. From the simulation results, the RAFT+ is capable of adapting to blockchain networks with IoT end-devices within the order of 10^2.

Figure 4.7 shows the probability of selecting different strong end-devices as the leader under different numbers of strong end-devices. Numbers on the histograms represent the serial numbers of strong end-devices as the leader. It can be seen that RAFT+ is able to select the optimal leader in the blockchain network according to the computing resources of IoT end-devices and the time-variant communication environment of the system.

4.5.3 Fault-Tolerant Performance Analysis

In the RAFT algorithm, the leader is responsible for block generation and dominating block consensus. After a new block is generated, the leader sends the new block to the followers for endorsement to reach a consensus on the new block. The followers then send confirmation responses containing the endorsement results to the leader. When the leader receives a sufficient number of confirmation responses,

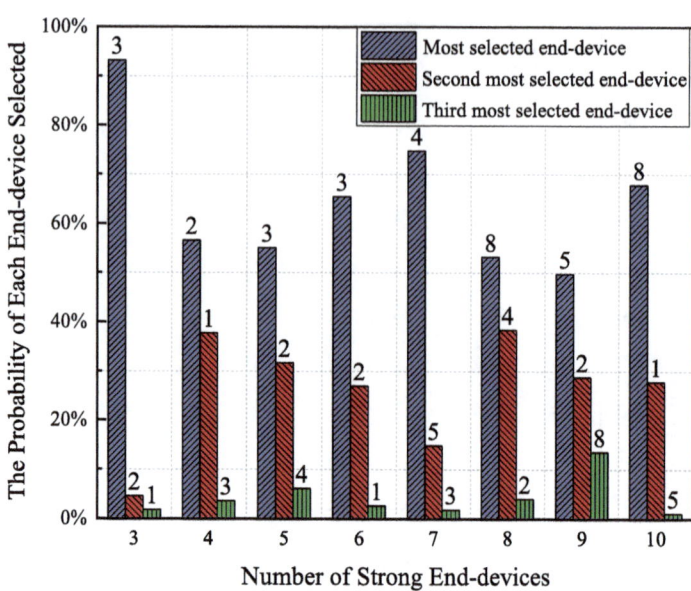

Fig. 4.7 The possibility of selecting different end-devices as the leader with different numbers of strong end-devices

an agreement has been achieved for the new block within the blockchain network. A synchronous block update message is reported to the followers by the leader. In the RAFT algorithm, leader selection is triggered by the heartbeat mechanism. If any follower fails to receive the heartbeat packet reported by the leader as scheduled, leader selection will be triggered. Meanwhile, the RAFT algorithm provides a data synchronization mechanism controlled by the leader to solve the problem of block inconsistency between different nodes with less computation required.

4.5.3.1 Block Inconsistency

In the blockchain network with the RAFT algorithm as the consensus mechanism, the blocks between nodes may not be synchronized, as shown in Fig. 4.8. The number in each block indicates the term of the block.

RAFT+ modifies the trigger conditions of leader selection and combines leader selection with block consensus. The leader in the blockchain network sends the new block to the followers to maintain the node state of the followers. The followers endorse the block and return confirmation responses including the endorsement results. If the leader receives more than $\lceil I/2 \rceil$ confirmation responses reported by the followers, an agreement is reached for the generation of the block in the network. During the block consensus process, a follower's failure to return confirmation response means the communication interruption occurs between the follower and

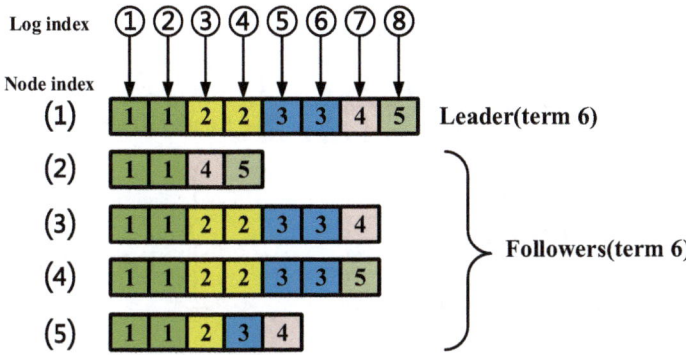

Fig. 4.8 Example of block inconsistency

the leader, and the follower is not able to update the block synchronously. During the operation of the system, the followers receive new blocks sent by the leader and return confirmation responses containing the endorsement results. After receiving a confirmation response, the leader checks the related data contained in the confirmation response. The data synchronization mechanism is triggered if the leader finds that the index of the block contained in the confirmation response is inconsistent with the index stored by itself. To maintain data consistency between the leader and the follower, the leader traverses both ledgers to find the closest consistent block. After the closest consistent block is found, the leader sends a command to delete all blocks after that block in the follower's ledger. Then, all blocks in the leader's ledger after that block are sent to the follower to complete the data synchronization.

In RAFT+, leader selection is triggered by hardware failure or communication failure. Communication failure means that more than $\lceil I/2 \rceil$ followers and the leader have communication interruptions. After leader selection is triggered, the central node in the blockchain network collects the parameters of each follower to select a new leader. The new leader collects the ledger of each follower and compares it with its ledger. The different parts of the follower's ledger will be covered according to the leader's ledger. Meanwhile, each follower reports the data that have been reported but not stored in the ledger to the new leader. As a part of block consensus, the new leader generates a block and sends it to the followers. After block consensus is approved, the followers update the block synchronously at the time determined by the new leader. Therefore, RAFT+ can solve the problem of block asynchronization among nodes that may occur in the blockchain network.

4.5.3.2   Network Fragmentation

During the operation of an IoT–blockchain system using the RAFT algorithm as the consensus mechanism, it may occur that several followers cannot receive

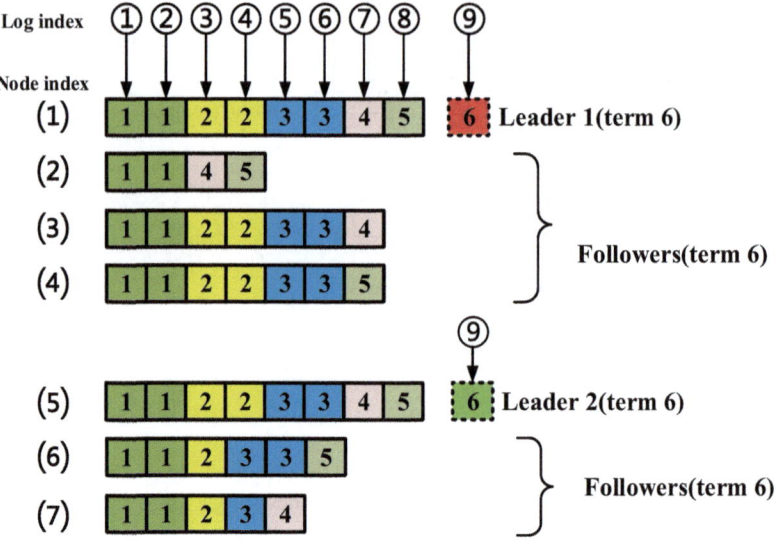

Fig. 4.9 Example of independent block generation after network fragmentation

the heartbeat packets at the same time and are converted to candidates. Several candidates send vote requests in parallel to compete for the votes of the followers in the network. The followers then vote for the candidate whose vote request they receive first (the candidate shall meet the requirements of selection parameters, such as term). Each candidate may have a random chance of gaining votes from the followers due to factors like communication environment and distance between nodes. If no candidate receives more than half of the votes, the vote will be invalid and leader selection will be restarted. In this case, the nodes in the blockchain network cannot work normally until a new leader is selected. The RAFT algorithm stipulates that only one leader can exist in a network to avoid network fragmentation. An example of network fragmentation is shown in Fig. 4.9, where the simultaneous existence of multiple leaders directly results in a split of the network and the independent generation of blocks. After the original network is split into multiple networks, each network will be independent of the others.

The blockchain networks require data consistency between nodes, and the block asynchronization caused by network fragmentation is unacceptable. To deal with the problem of network fragmentation in the blockchain networks, RAFT+ modifies the process that several candidates send vote requests in parallel to compete for the votes of the followers in leader selection. Firstly, RAFT+ determines whether to select a new leader based on the failure situation of the original leader. Secondly, RAFT+ considers all followers as candidates for leader selection, so that the optimal leader choice can be made in each term. Finally, RAFT+ selects a central node in the blockchain network to collect the parameters of each follower. The central node selects the unique leader by combining the original leader selection parameters in the RAFT algorithm with the unique parameters under the IoT application scenarios, such as the computing resources of end-devices and time-variant communication environment of the system. This kind of leader selection method can effectively avoid the existence of multiple leaders in the network at the same time, thus preventing network fragmentation.

Based on the above analysis, it can be seen that RAFT+ proposed in this chapter can effectively avoid network fragmentation and maintain the fault-tolerant performance of the original RAFT algorithm.

4.6 Summary

In this chapter, RAFT+ was provided as the consensus mechanism of the BIoT system, which can significantly reduce the communication loss and computing load caused by block consensus. Moreover, RAFT+ incorporated a DQN-based leader selection scheme, which inherits the workflow of the original leader selection mechanism. Meanwhile, the leader selection scheme improved the fault tolerance performance of the blockchain network and effectively avoided network fragmentation. A performance evaluation was conducted based on simulations to eliminate unstable factors during system operation. Simulation results showed that the average block generation latency of RAFT+ was reduced by 10% to 21% compared with the original RAFT algorithm under different conditions. It means that RAFT+ is able to complete block consensus with low latency, so as to improve the efficiency of the system and maintain its stability under high load conditions. In the next chapter, a BIoT system in real scenarios is implemented to verify the effectiveness.

References

1. S. Pahlajani, A. Kshirsagar, V. Pachghare, Survey on private Blockchain consensus algorithms, in *2019 1st International Conference on Innovations in Information and Communication Technology (ICIICT)* (2019), pp. 1–6
2. Y. Xiao, N. Zhang, W. Lou, Y.T. Hou, A survey of distributed consensus protocols for Blockchain networks. IEEE Commun. Surv. Tutorials **22**(2), 1432–1465 (2020)
3. D. Ongaro, J. Ousterhout, In search of an understandable consensus algorithm, in *2014 USENIX Annual Technical Conference (USENIX ATC 14)* (2014), pp. 305–319
4. L. Hou, X. Xu, K. Zheng, X. Wang, An intelligent transaction migration scheme for RAFT-based private Blockchain in Internet of Things applications. IEEE Commun. Lett. **25**(8), 2753–2757 (2021)
5. P. Danzi, A.E. Kalør, Č. Stefanović, P. Popovski, Delay and communication tradeoffs for Blockchain systems with lightweight IoT clients. IEEE Internet Things J. **6**(2), 2354–2365 (2019)
6. G. Fortino, C. Savaglio, G. Spezzano, M. Zhou, Internet of Things as system of systems: a review of methodologies, frameworks, platforms, and tools. IEEE Trans. Syst. Man Cybern.: Syst. **51**(1), 223–236 (2021)

7. L. Lei, Y. Tan, K. Zheng, S. Liu, K. Zhang, X. Shen, Deep reinforcement learning for autonomous internet of things: model, applications and challenges. IEEE Commun. Surv. Tutorials **22**(3), 1722–1760 (2020)
8. Q. Zhou, K. Zheng, L. Hou, J. Xing, R. Xu, Design and implementation of open LoRa for IoT. IEEE Access **7**, 100649–100657 (2019)
9. X. Xiong, K. Zheng, L. Lei, L. Hou, Resource allocation based on deep reinforcement learning in IoT edge computing. IEEE J. Sel. Areas Commun. **38**(6), 1133–1146 (2020)
10. W. Iqbal, H. Abbas, M. Daneshmand, B. Rauf, Y. A. Bangash, An in-depth analysis of IoT security requirements, challenges, and their countermeasures via software-defined security. IEEE Internet Things J. **7**(10), 10250–10276 (2020)
11. Z. Liu, Q. Zhou, L. Hou, R. Xu, K. Zheng, Design and implementation on a LoRa system with edge computing, in *2020 IEEE Wireless Communications and Networking Conference (WCNC)* (2020), pp. 1–6
12. S. Guo, X. Hu, S. Guo, X. Qiu, F. Qi, Blockchain meets edge computing: a distributed and trusted authentication system. IEEE Trans. Ind. Inf. **16**(3), 1972–1983 (2020)
13. Z. Yang, K. Yang, L. Lei, K. Zheng, V.C.M. Leung, Blockchain-based decentralized trust management in vehicular networks. IEEE Internet Things J. **6**(2), 1495–1505 (2019)
14. S. Malik, V. Dedeoglu, S.S. Kanhere, R. Jurdak, Trustchain: trust management in Blockchain and IoT supported supply chains, in *2019 IEEE International Conference on Blockchain (Blockchain)* (2019), pp. 184–193
15. Y. Yu, S. Liu, P. Yeoh, B. Vucetic, Y. Li, LayerChain: a hierarchical edge-cloud blockchain for large-scale low-delay IIoT applications. IEEE Trans. Ind. Inf. **3203**(c), 1–1 (2020)
16. B. Li, Y. Wu, J. Song, R. Lu, T. Li, L. Zhao, Deepfed: federated deep learning for intrusion detection in industrial cyber–physical systems. IEEE Trans. Ind. Inf. **17**(8), 5615–5624 (2021)
17. G. Wang, Z. Shi, M. Nixon, S. Han, Chainsplitter: towards Blockchain-based industrial IoT architecture for supporting hierarchical storage, in *2019 IEEE International Conference on Blockchain (Blockchain)* (2019), pp. 166–175

Chapter 5
Prototype Implementation of BIoT

5.1 Introduction

In this chapter, we present the design and prototype implementation of a BIoT system in a real-world scenario to demonstrate its feasibility. The BIoT prototype, named HyperLoRa, is based on the LoRa system with edge computing, exemplifying typical IoT systems. The blockchain component is implemented using the open-source Hyperledger Fabric. Following the architecture outlined in Chap. 3, HyperLoRa incorporates a multiple ledger design and integrates transaction and block construction into the LoRa system's procedures.

Existing designs of blockchain-enabled LoRa systems can be generally categorized into two types. The first type involves deploying the blockchain solely in the central cloud, e.g., as seen in [1, 2]. While this approach enhances data security through blockchain, it fails to leverage the edge computing capabilities of LoRa gateways. Additionally, it increases the workload on the central cloud, which must also maintain the blockchain. The second type involves directly implementing the blockchain at LoRa gateways without a specific design, which lacks implementation feasibility [3]. As the data of blockchain grow, the limited computing and storage resources of LoRa gateways may quickly become exhausted. Meanwhile, the blockchain with decentralized ledgers can help to protect LoRa data from being falsified when it is applied in LoRa systems. To address these issues, a blockchain-enabled LoRa system with edge computing is expected to be used. This system aims to enhance the performance and security of LoRa systems by utilizing the computing capabilities of LoRa gateways more effectively.

Each node in the blockchain network maintains a complete copy of the ledger, which records transactions in chained blocks. In the provided system, two distinct ledgers are constructed to process and store different types of data, i.e., delay-tolerant application data of large size are stored in a ledger located in the central cloud, while time-critical network data of smaller size are stored in a ledger at the LoRa gateways. This setup effectively leverages the edge computing capabilities

© The Author(s), under exclusive license to Springer Nature Switzerland AG 2024

L. Hou et al., *Blockchain-Based Internet of Things*, SpringerBriefs in Computer
Science, https://doi.org/10.1007/978-3-031-70303-4_5

of LoRa gateways. The provided system allows two main features of the network servers, i.e., join procedure handling and application packages processing, to be migrated to the LoRa gateways. Consequently, this migration helps balance the workload between the central cloud and the LoRa gateways. As a result, the central cloud's CPU utilization is optimized, and the network bandwidth required between the LoRa gateways and the central cloud is reduced.

Moreover, a HyperLoRa prototype is also implemented on embedded hardware to demonstrate the feasibility of our provided design. To run Hyperledger Fabric on LoRa gateways, which are built on ARM64 architecture CPUs, the embedded Linux operating system has been customized to be compatible with a third-party Docker image of Fabric. Two channels have been configured and governed by two organizations, i.e., the LoRa Gateway Organization and the Network Server Organization. Each channel consists of an ordering node for consensus and multiple peers that maintain a shared ledger. LoRa gateways can request and query transactions by invoking chaincodes via the Fabric client. Additionally, the join server (JS) and network connector (NC) modules are implemented at LoRa gateways, utilizing their edge computing capabilities. The prototype implementation of HyperLoRa provides the viability of integrating blockchain into LoRa gateways, despite the resource constraints of embedded hardware. Experimental results show that HyperLoRa achieves equivalent performance with reduced resource consumption compared to traditional LoRa systems.

5.2 Related Works

Efficiency and security issues become critical for the development of IoT services. As a promising solution, blockchain has attracted lots of attention, since it can ensure the data security and privacy in a distributed manner [4, 5] or provide a secure way to trade items [6]. Besides, edge computing can provide efficient and decentralized data services, such as data analytics or distributed machine learning at the edge, because it is closer to the data source than central cloud [7, 8]. However, due to the various requirements and the limited resources in IoT devices, it is still challenging on how to integrate blockchain into different IoT scenarios both theoretically and technically.

5.2.1 Implementations of LoRa Systems

There are several implementations of LoRa system, e.g., the XisLoRa,[1] the Chirp-Stack,[2] and the Things Network (TTN).[3] Their designs are all based on the reference

---

[1] https://github.com/xisiot/lora-system.

[2] https://www.chirpstack.io/.

[3] https://www.thethingsnetwork.org/.

model that is proposed in [9]. XisLoRa separates a network connection module to handle the package parsing and verification, while ChirpStack uses a gateway bridge to communicate directly with LoRa gateways. TTN follows the reference model and deploys one network server to handle all kinds of requests. These implementations have two things in common. On the one hand, all the LoRa gateways in these systems are only responsible for package forwarding. No further work is required for LoRa gateways to finish. Thus, the resources at LoRa gateways are not efficiently utilized. On the other hand, all the data in these systems are stored in traditional databases, such as MySQL (by XisLoRa), PostgreSQL (by ChirpStack), and InfluxDB (by TTN). Therefore, these systems have potential risks of data falsification and leakage without blockchain.

5.2.2   Blockchain-Enabled LoRa Systems

As one of the promising IoT transmission techniques, LoRa is widely used for scenarios with low-power and wide-area requirements. Blockchain is suitable for a LoRa system to ensure data security. However, it is challenging to integrate blockchain into LoRa system. This is because most of the end-devices are resource-constrained. They cannot maintain ledgers in blockchain network. Meanwhile, the LoRa gateways are not as capable as central cloud; they are only used for transparent transmission of LoRa packages between end-devices and central cloud. Deploying blockchain in LoRa gateways needs a redesign of LoRa system architecture and well consideration of the resources. Liu et al. [10] propose an edge computing-based LoRa system in order to migrate the functions of rejoin and media access control (MAC) commands into LoRa gateways to mitigate the workloads of central cloud. However, databases are deployed in LoRa gateways to store data, which requires an extra mechanism for data synchronization. Both Lin et al. [2] and Danish et al. [1] propose their designs for LoRa systems where blockchain is only deployed in central cloud. These designs cannot ease the burden of central cloud but entail more workloads on it. The edge computing abilities of LoRa gateways are not enabled. Ozyilmaz et al. [3] propose a blockchain-based LoRa system where entities with different computing and storage abilities play different roles. LoRa gateways with high abilities can download full blockchain, while weak LoRa gateways can only download the headers of blocks. However, such a design has ignored that the increasing size of the ledgers in the blockchain may exceed the storage limitation of LoRa gateways. Besides, the intention of deploying blockchain in LoRa gateways is to verify blocks and store data, making no benefit for the efficiency of the whole system.

5.3 Overview of LoRa System

5.3.1 Network Reference Model

Most of the common LoRa systems are built based on the network reference model that is recommended in [9]. The specification suggests that the LoRa system should contain five basic parts, i.e., end-device, LoRa gateway, network server, join server, and application server, which is illustrated in Fig. 5.1. The end-devices are applied for IoT applications, such as sensing the environment. The application data are transmitted from end-devices to LoRa gateways via LoRa wireless links. The network servers, join servers, and application servers are deployed in central cloud. LoRa gateways are responsible for package forwarding between end-devices and network servers. Then, all the application messages are processed by network servers. Among all these messages, the join messages are transferred to join servers, while the application messages are sent to application servers. All data are created, read, updated, or deleted in databases.

5.3.2 Basic Features and Categories of LoRa Data

The LoRa system mainly accomplishes two functions, i.e., the authentication of end-devices via join procedure and the transmission of application data between end-devices and application servers.

Fig. 5.1 Illustration of LoRa network reference model

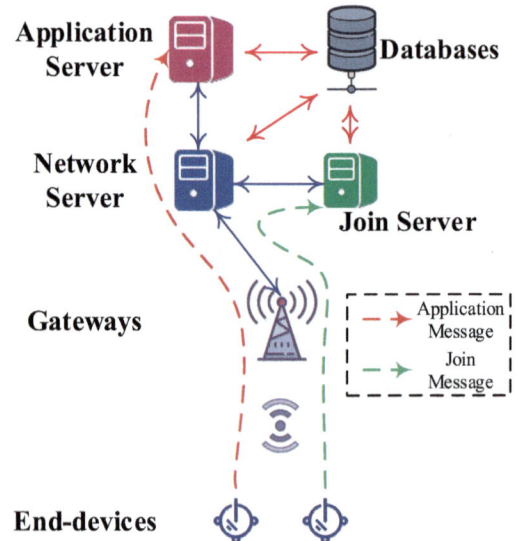

1. **Join procedure:** A join procedure is required for an end-device to be activated and access the LoRa network. There are two modes of join procedure, i.e., over-the-air activation (OTAA) and activation by personalization (ABP). OTAA works by dynamic negotiations between end-device and join server, while in ABP mode, all the session context are preset manually and remain unchanged in both end-device and join server. Either way, a session to exchange messages is established between end-device and network server.

2. **Transmission of application data:** The application data are wrapped into LoRaWAN packages by end-devices for uplink transmission or by application servers for downlink transmission. Therefore, the network servers need to unpack/pack the packages, authenticate/specify the end-devices, and verify/ensure the integrity of the packages for uplink/downlink transmission of application data.

Basically, the data that are involved in the two functions can be classified into two categories, i.e., the session context data of end-devices and the application data. The context data contain all the necessary information to keep connections between end-devices and network servers, for example, the end-device address (DevAddr), the Application Session Key (AppSKey) for encryption/decryption of application data and the Network Session Key (NwkSKey) for calculation of message integrity code (MIC) value, and some configurations of LoRa physical layer. The application data are application-specific and owned by application providers or end-users. Take the environment monitoring as an example, and the application data are the environment data that are gathered from end-devices with sensors.

5.4 Design of HyperLoRa System

5.4.1 Architecture with Multiple Ledgers

The set of LoRa gateways and network servers are given as $\mathscr{G} = \{G_1, G_2, \ldots, G_K\}$ and $\mathscr{S} = \{S_1, S_2, \ldots, S_M\}$, respectively. G_k stands for the GatewayEUI of the k-th gateway that is used to identify LoRa gateway, while S_m represents the identity (ID) of network server m. Each LoRa gateway or network server has a pair of secret keys which are denoted as $< k_\varepsilon^{pub}, k_\varepsilon^{pri} >$, where $\varepsilon \in \mathscr{G} \cup \mathscr{S}$. k_ε^{pub} is the public key that is freely accessed by all other entities, while k_ε^{pri} is the private key that is only possessed by the owner. As illustrated in Fig. 5.2, HyperLoRa consists of a network ledger $\mathscr{N}$ and an application ledger $\mathscr{A}$. All data are stored as immutable records of transactions in ledgers. The main mathematic notations are listed in Table 5.1. The functions of the two ledgers are elaborated as follows, from perspectives of both data types and node types:

- **Perspective of data types:** All session context of end-devices are kept in $\mathscr{N}$. These data include all the attributes of end-devices that remain unchanged during

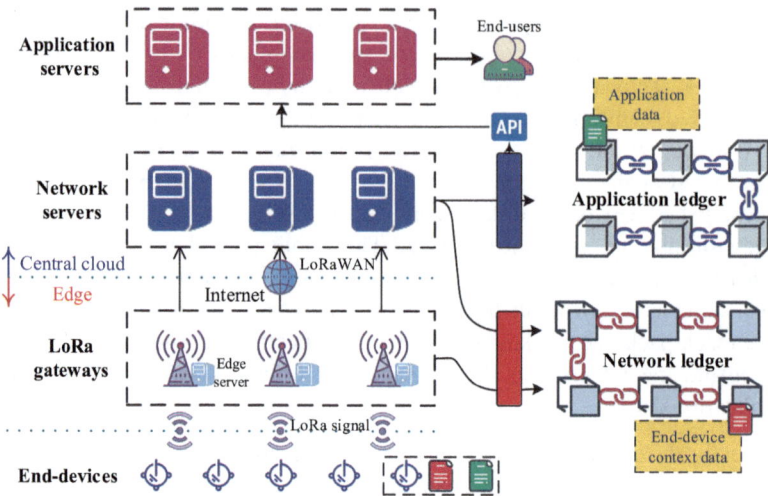

Fig. 5.2 Illustration of HyperLoRa system architecture

Table 5.1 List of mathematical notations

Notation	Explanation	Notation	Explanation
$\mathscr{G}$	Set of LoRa gateways	$\mathscr{S}$	Set of network server
ε	*An entity of LoRa gateway or network server	k_ε^{pub}	Public key of entity ε
k_ε^{pri}	Private key of entity ε	$\mathscr{N}$	Network ledger
$\mathscr{A}$	Application ledger	B	Block
T	Transaction	h	Message digest of transaction
t	Request time of transaction	d	Raw data in transaction
τ	Generation timestamp of block	$\mathscr{M}$	Root value of Merkle tree
$\mathscr{H}$	Hash value of previous block		

one session. The Device Extended Unique Identifier (DevEUI) field is created when the end-device registers to join servers, while the other fields are generated during the join procedure. The application data that are produced by end-devices are stored in $\mathscr{A}$. Only application servers are authorized to access these data for further processing.

- **Perspective of node types:** $\mathscr{N}$ is maintained by all LoRa gateways and network servers. LoRa gateways can generate new blocks for $\mathscr{N}$ in OTAA mode, while network servers create new blocks for end-devices in ABP mode. This is because LoRa gateways receive join requests from OTAA mode end-devices directly. Thus, they can pack session context data into blocks without wasting time and resources forwarding data to network servers. Both LoRa gateways and network servers are responsible for block verification and synchronizing on their ledgers. As for $\mathscr{A}$, the new blocks are generated and validated only by network servers.

However, some application programming interfaces (APIs) are provided for the development of application servers.

The session context data and the application data have different characteristics and purposes. The session context data are used to authenticate end-devices, encrypt messages, and calculate MIC. The amount of session context data is small, and the changing frequency is low. Therefore, LoRa gateways can be able to maintain $\mathcal{N}$, because there are extra resources apart from those that are used by physical layer. Different from that, the size of $\mathcal{A}$ can be increasing quickly. Besides, application data should only be accessed by application servers to provide data services for end-users. Thus, only network servers can maintain $\mathcal{A}$. Although all the data are generated by end-devices originally, they cannot participate in blockchain network due to their limitation on resources.

5.4.1.1   Network Ledger

The context data are recorded in $\mathcal{N}$ in the form of transactions which is denoted as $T^{\mathcal{N}}$. The content of $T^{\mathcal{N}}$ is given as

$$T^{\mathcal{N}} = < \varepsilon, h, t, \bar{d} >, \tag{5.1}$$

where h is the message digest of the transaction, t is the timestamp of the transaction, and $\bar{d}$ is the session context data after encryption. The raw d is given as

$$d = \text{DevEUI} \parallel \text{AppKey} \parallel \text{DevAddr} \parallel \text{NwkSKey} \parallel \text{Nonces},$$

where operator $\parallel$ means the concatenation of the data, and

$$\text{Nonces} = \text{DevNonce} \parallel \text{AppNonce}.$$

Note that the AppSKey is not included because it is related to applications and should only be kept by application servers. Each LoRa gateway needs to encrypt d and calculate h to generate a $T^{\mathcal{N}}$ using the key pairs. $\bar{d}$ is calculated as

$$\bar{d} = \text{Enc}\{k_{\varepsilon}^{\text{pub}}, d\}, \tag{5.2}$$

and h is derived as

$$h = \text{Sign}\{k_{\varepsilon}^{\text{pri}}, t \parallel \bar{d}\}, \tag{5.3}$$

where operator Enc stands for an encryption algorithm using the public key of the requester and Sign represents a digital signature algorithm with private key. Therefore, d can only be decrypted out using the corresponding private key, and

---

Algorithm 3: The generation of Merkle tree and root value

Input: All transactions $\{T_1^{\mathscr{N}}, T_2^{\mathscr{N}}, \ldots, T_R^{c}\}$, let $R = 2^x + y$, where $x \geq 0, 0 \leq y < 2^x$ and
  $z = 1$
Output: The root hash $\mathscr{M}$

for z *in* $1, 2, \ldots, x + 1$ **do**
 for i *in* $1, 2, \ldots, \lceil \frac{R}{2^z} \rceil$ **do**
  if $z = 1$ **then**
   if $2i \leq R$ **then**

$$hash_{z,i} = H(h_{2i-1} || h_{2i}),$$

    where H is a hash function.
   end
  else

$$hash_{z,i} = h_{2i-1}$$

   end
  end
  else
   if $2i \leq \lceil \frac{R}{2^{z-1}} \rceil$ **then**

$$hash_{z,i} = H(hash_{z-1,2i-1} || hash_{z-1,2i}),$$

   end
   else

$$hash_{z,i} = hash_{z-1,2i-1}$$

   end
  end
 end
end
$\mathscr{M} = hash_{x+1,1}.$

---

h can be verified by any LoRa gateway or network server that has the public key of the requester.

In $\mathscr{N}$, after reaching a predefined condition (e.g., a period of time or maximum number of transactions), the LoRa gateway starts to generate a new block that contains all the transactions after that period. Assuming that totally R transactions need to be packaged at time τ and letting $T_r^{\mathscr{N}} = < \varepsilon_r, h_r, t_r, \bar{d}_r >$ denote the r-th transaction, the LoRa gateway needs to construct a Merkle tree before generating the block according to Algorithm 3.

After creating the Merkle tree and getting the root value $\mathscr{M}$, LoRa gateway starts to generate a block $B^{\mathscr{N}}$ which includes an index indicator $\zeta \in \mathbb{Z}$, a body $b^{\mathscr{N}}$, and a header $e^{\mathscr{N}}$. $b^{\mathscr{N}}$ is given as follows:

$$b^{\mathcal{N}} = \varepsilon_1||h_1||t_1||\bar{d}_1||\varepsilon_2||h_2||t_2||\bar{d}_2||\ldots||\varepsilon_R||h_R||t_R||\bar{d}_R.$$

Keeping the ε field in blocks is necessary, because transactions can be transferred from one LoRa gateway to others. The $e^{\mathcal{N}}$ is given as $e^c = <\tau, \mathcal{M}, \mathcal{H}>$, where τ is the timestamp of the generation of the block, $\mathcal{M}$ is the root of Merkle tree, and $\mathcal{H}$ is the hash value of the previous block. For the i-th block, the $\mathcal{H}_i$ is derived as follows:

$$\mathcal{H}_i = H\left(\zeta_{i-1}||e^{\mathcal{N}}_{i-1}||b^{\mathcal{N}}_{i-1}\right).$$

The new block is broadcast to all other LoRa gateways and network servers in $\mathcal{N}$. Others need to verify the block and endorse it if it is considered valid. At first, other nodes need to check h of all $T^{\mathcal{N}}$ to make sure they come from exactly a trustful LoRa gateway. With the ID value, the $k^{\text{pub}}_{\varepsilon}$ can be derived, and h can be verified using the same algorithm as the signing procedure. Then, the root of Merkle tree is calculated again to ensure its validity. Note that the session context data of end-devices are encrypted; thus, the data cannot be eavesdropped during transmission, and other nodes need not to verify the raw content of transactions.

After verification, other nodes need to broadcast a signed result. According to PBFT, if each node receives more than $2p + 1$ (where p is the maximum possible node that is not functioning well) results that are the same, then the consensus is reached. All nodes synchronize $B^{\mathcal{N}}$ on their ledger.

5.4.1.2 Application Ledger

The application data are stored in $\mathcal{A}$ as transactions, which is denoted as $T^{\mathcal{A}}$. The structure of $T^{\mathcal{A}}$ is similar to $T^{\mathcal{N}}$ except for the identifier of requester and the data field. Let $T^{\mathcal{A}} = <\varepsilon, h, t, d>$, where $\varepsilon \in \mathcal{S}$ and d is the application data. Note that the requester of $T^{\mathcal{A}}$ can only be network servers, and the data in transactions need not to be deciphered by network servers, because the application data have been encrypted by end-devices using AppSKey and kept encrypted during the processing.

When a network server receives enough $T^{\mathcal{A}}$, it starts to wrap them into a new block. By reaching consensus with all other network servers, the new block can be recorded on $\mathcal{A}$, and the application data d are finally stored on the shared ledger. End-users can access the application data through application servers that could get application data from APIs and decrypt them using proper AppSKey, as shown in Table 5.2.

5.4.2 Edge Computing in LoRa Gateways

With edge computing abilities in HyperLoRa, two modules are moved from network servers to LoRa gateways, i.e., JS that handles the join procedures of end-devices

Table 5.2 Information of each component in HyperLoRa system

Component	Functions	Capability of computing and storage	Ledger
End-device	Sensing, actuating, ...	Very low	None
LoRa gateway	Signal processing, data forwarding, join procedure handling, application packages processing	Mediocre	$\mathscr{N}$
Network server	Data analytics, application APIs, ...	High	$\mathscr{N}, \mathscr{A}$

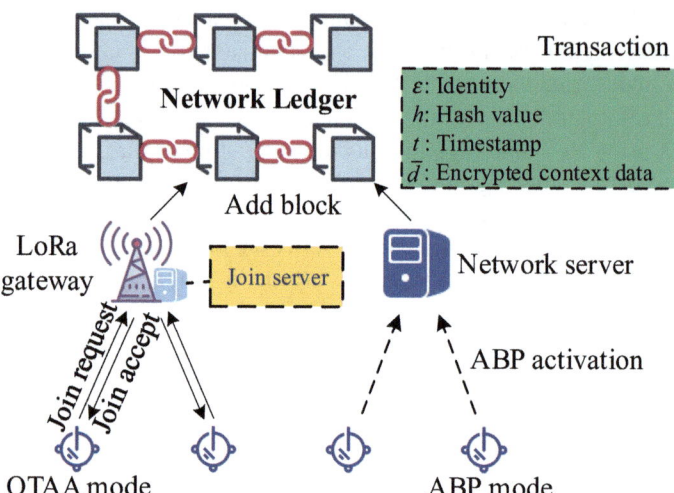

Fig. 5.3 Illustration of join procedure

and NC that is responsible for application package processing. By join procedures, the context data of end-devices are generated and stored in $\mathscr{N}$. On the other hand, LoRa gateways need these context data from $\mathscr{N}$ to fulfill the tasks of application package processing.

5.4.2.1   Join Procedure Handling

As shown in Fig. 5.3, the join procedure of OTAA mode is mainly handled by LoRa gateways, while the ABP mode activation of end-devices is processed by network servers.

- **OTAA:** As shown in Fig. 5.4, when a LoRa gateway receives a join request, it checks the validity of the request, including the MIC value. Then, LoRa gateway needs to generate session context data for this end-device, including

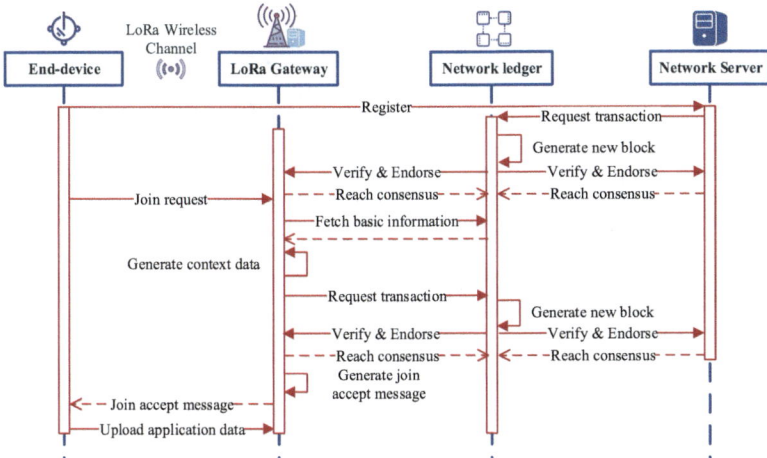

Fig. 5.4 Workflow of join procedure in OTAA mode

the DevAddr, two session keys, and some metadata. A transaction that contains these data is requested and put into a new block by the LoRa gateway. After consensus, the new block is added into $\mathcal{N}$. In the meantime, the LoRa gateway has to generate a join accept message and send it back to end-device. Successful reception of a join accept message by end-device indicates the success of the join procedure.

- **ABP:** In ABP mode, the manufacturers or owners of end-devices need to fill all session context data manually via the interfaces offered by network servers. Then, network servers request transactions of these data and add blocks on $\mathcal{N}$ after consensus. No more negotiations are needed between end-devices and JS in LoRa gateways before end-devices can upload application data.

5.4.2.2 Application Package Processing

Owing to $\mathcal{N}$ being deployed in LoRa gateways, an NC module that is used for application package processing is moved from network server to LoRa gateway, as shown in Fig. 5.5. The workflow of message exchange between end-device and the HyperLoRa system is depicted in Fig. 5.6.

- **Uplink package**
  When the end-device uploads an application package, the NC in the LoRa gateway parses the package into three parts: the metadata, the encrypted application data, and the MIC value. By extracting the DevAddr field from the metadata, the NC can query the session context data of the end-device from $\mathcal{N}$. It then checks the validity of the MIC value to ensure the integrity of the package. If

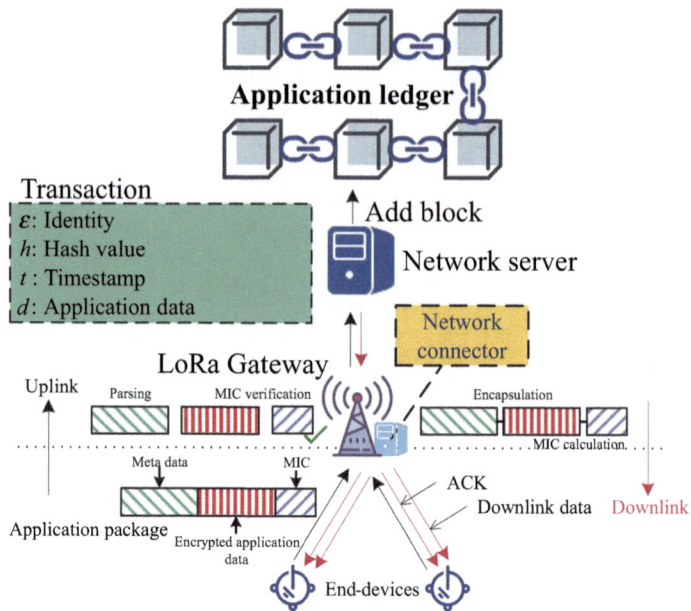

Fig. 5.5 Illustration of application package processing

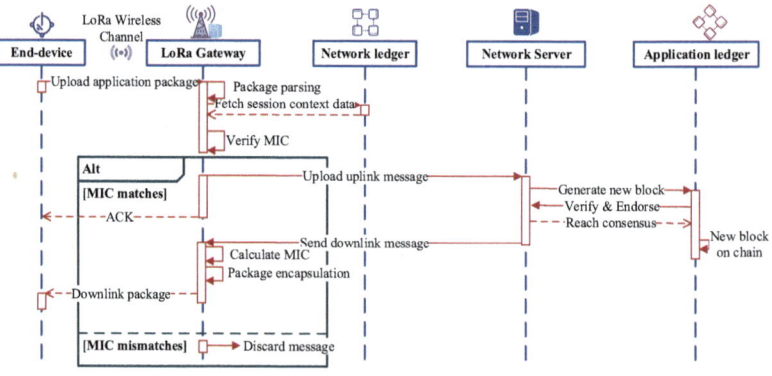

Fig. 5.6 Workflow of application package processing

the MIC value mismatches, the package is directly discarded. Otherwise, the LoRa gateway uploads the encrypted application data to the network server. The application data are packed into blocks that will be added to $\mathscr{A}$ by the network servers. Meanwhile, the LoRa gateway needs to inform the end-device of the successful reception of the package with an **ACK!** (ACK!) message.

- **Downlink package:**
  If there are any downlink data for end-devices, the network servers need to send the downlink data to the LoRa gateways simultaneously while adding a new

block to the chain. Only the encrypted data need to be sent from the network server to the LoRa gateways, rather than the full packages. The LoRa gateways then create and calculate the MIC value and encapsulate the downlink package with the necessary information in the NC. During this process, the context data of the end-devices are retrieved by $\mathcal{N}$. The downlink package is finally transmitted to the end-devices via LoRa wireless channels.

With edge computing that leverages the LoRa gateway near the end-devices for preliminary processing, the efficiency of central cloud in HyperLoRa can be improved. The tasks of package parsing, encapsulation, and MIC calculation are performed by the LoRa gateways at the edge, rather than by the network servers in the central cloud. This approach saves a significant amount of computing resources in the central cloud by utilizing the spare resources of the LoRa gateways. Additionally, invalid messages, such as those with incorrect MIC values, are blocked by the LoRa gateways, thereby preventing unnecessary consumption of computing resources in the central cloud and avoiding congestion in the transmission links between the LoRa gateways and the central cloud.

5.5 Security Analysis

In this section, we analyze the security enhancements provided by the HyperLoRa system concerning four types of risks, i.e., application layer Denial of Services (DoS) attacks, Single Point of Failure (SPOF) (Single Point of Failure), malicious LoRa gateways, and malicious network servers. The deployment of blockchain and edge computing in HyperLoRa enables effective mitigation or elimination of these risks.

5.5.1 DoS Attack on Application Layer

For traditional LoRa system, the interfaces between network servers and LoRa gateways are defined in [11]. Therefore, an attacker can launch DoS attack by sending a large number of invalid application packages to network servers in high frequency to exhaust computing and bandwidth resources. In HyperLoRa system, the application packages are processed in LoRa gateways, and the interface protocol between network servers and LoRa gateways can be defined privately. It could take attackers much more effort to crack the private protocol and launch DoS attack toward network servers. It is also very hard to attack HyperLoRa through LoRa gateways, because they are geographically distributed and they only receive physical LoRa signals. LoRa gateways can also filter malicious traffic for network servers; thus, it is guaranteed that network servers can stay healthy to ensure the functions of the HyperLoRa system during the DoS attack through LoRa gateways.

5.5.2   Single Point of Failure

In HyperLoRa, the SPOF issue for both LoRa gateways and network servers is mitigated because each LoRa gateway and network server maintains a full copy of the shared ledger. If a LoRa gateway malfunctions, nearby gateways can take over its functions as long as they can receive signals from the end-devices. They only need to securely obtain the private key of the malfunctioning gateway. Similarly, if a network server fails, other network servers can take its place, and LoRa gateways can re-upload data to the alternative servers.

Additionally, by migrating the Join Server (JS) and Network Controller (NC) functions into the LoRa gateways, the failure of a single network server does not affect the handling of join procedures and application package processing. This migration, coupled with the edge computing capabilities of the LoRa gateways, helps balance the workload between the network servers and the gateways. As a result, the likelihood of network server crashes is significantly reduced.

5.5.3   Malicious LoRa Gateway

A malicious LoRa gateway can do two things to HyperLoRa system. The first kind is to steal session context data of end-devices from $\mathcal{N}$. The other kind of thing is to disturb the consensus process. In both ways, the malicious LoRa gateway needs to pass the authentication process and participate in the blockchain in HyperLoRa. If the shared ledger is retrieved, the malicious LoRa gateway also needs to obtain the k^{pri} of other LoRa gateways to decrypt the session context data in blocks. This could be extremely difficult because LoRa gateways can keep the k^{pri} in specialized encryption chips. For consensus process, attackers need to crack more than p LoRa gateways in HyperLoRa to break the PBFT algorithm, which is difficult to realize.

5.5.4   Malicious Network Server

Assume an attacker has deployed in HyperLoRa system a malicious network server that has successfully passed the authentication and has established connections with some LoRa gateways (which is actually difficult to realize). Then, the malicious network server can share both $\mathcal{N}$ and $\mathcal{A}$. However, neither the session context data of end-devices nor the application data can be stolen, falsified, or destroyed. First of all, all the session context data are encrypted using k^{pri}, while all the application data are encrypted using AppSKey that is kept only by end-devices and application servers. To decipher the data, the attacker needs also to get all these keys. Then, the blockchain in HyperLoRa ensures that the data inside the blocks cannot be altered. Moreover, all nodes share a full copy of ledgers; thus, deleting data from one network server will not damage the data that are kept in other LoRa gateways or network servers.

5.6 Experimental Results and Analysis

5.6.1 Implementation with Hyperledger Fabric

The prototype of the HyperLoRa system is implemented using Hyperledger Fabric for the blockchain, as illustrated in Fig. 5.7. The end-device is equipped with a Semtech SX1276 LoRa transceiver module. By connecting an appropriate sensor, the end-device can collect data and upload it to the LoRa gateway. The LoRa gateway is built on embedded hardware featuring a Smart6818 CPU board, which is industrially specialized and utilizes a Samsung S5P6818 octa-core Cortex-A53 System on Chip (SoC). This hardware enables edge computing capabilities. Additionally, two radio frequency front-ends (SX1255) are deployed and connected to a digital baseband chip (SX1301) for signal processing.

To deploy Hyperledger Fabric on the LoRa gateway, the operating system is customized based on the Linux OpenWrt project, as Fabric requires the cgroup feature and an overlay file system. Additionally, a third-party Docker image of Fabric is built to run on the LoRa gateway.

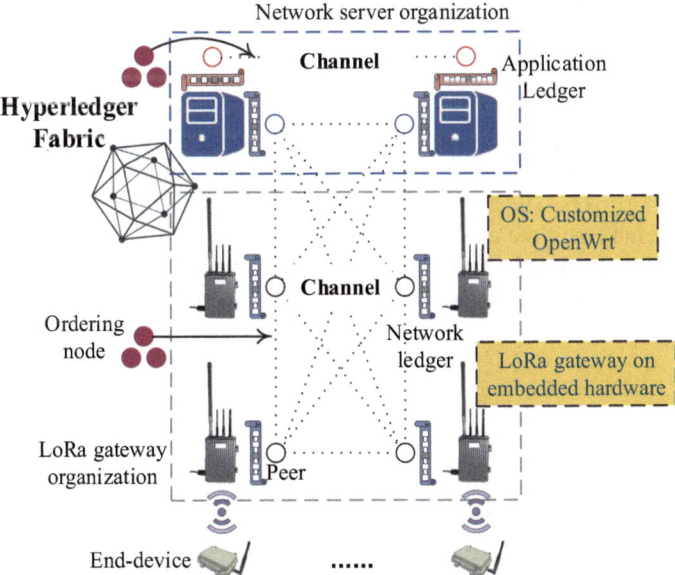

Fig. 5.7 Implementation of HyperLoRa with embedded LoRa gateways and Hyperledger Fabric

During the initialization of Fabric, two organizations are created, i.e., the LoRa gateway organization, which contains peers deployed in LoRa gateways, and the network server organization, which includes peers deployed in network servers. Each peer functions as both an endorser and a committer. Two channels are established, each maintaining a separate ledger, i.e., $\mathcal{N}$ and $\mathcal{A}$. All peers in both organizations can communicate within the $\mathcal{N}$ channel, while only peers in the network server organization can communicate privately within the $\mathcal{A}$ channel.

As a permissioned blockchain, a Membership Service Provider (MSP) is established to manage the authentication and authorization of all peers, ensuring that only authorized LoRa gateways or network servers can join the blockchain network. The consensus is achieved using the PBFT algorithm, with several ordering nodes in each channel. Due to the limited computational capabilities of LoRa gateways, all ordering nodes for both channels are deployed in the central cloud.

Several chaincodes are developed and hosted on LoRa gateway peers to manipulate data in the $\mathcal{N}$ ledger. Both the JS and NC can invoke these chaincodes via a Fabric client to create, read, or delete session context data of end-devices. Peers run the chaincodes in a virtual environment to validate them. Once peers respond, the JS or NC sends the transaction to the ordering nodes for consensus. After achieving consensus, the data are recorded on all distributed instances of $\mathcal{N}$, and the latest values of specific data fields are maintained as the world states of $\mathcal{N}$ in a state database.

5.6.2 Experimental Setup

Our prototype system consists of four LoRa gateways and two network servers at the central cloud. To emulate hundreds of end-devices, several customized Locust[4] clients are deployed on independent servers that can connect to LoRa gateways directly. A Fabric network of version 1.4.0 is established on both LoRa gateways and network servers. The hardware information is summarized in Table 5.3.

Key parameters of the experimental system are listed in Table 5.4. The ordering type is *solo*, which means the consensus is reached by a single ordering node using sorting algorithm. New block is generated every 2 s. If the total number of

Table 5.3 Hardware information

Type of device	Processor	Number of CPU cores	Memory
LoRa gateway	Samsung(R) S5P6818 @ 1.40 GHz	8	1 GB
Network server	Intel(R) Xeon(R) W-2123 @ 3.60 GHz	8	32 GB
Locust client server	Intel(R) Xeon(R) CPU E5-2609 v4 @ 1.70 GHz	8	16 GB

---

[4] https://www.locust.io/.

Table 5.4 Parameter settings

Name	Value	Description
General settings		
OrdererType	Solo	Type of order in Hyperledger
BatchTimeout	2 s	Time interval to generate a block
MaxMessageCount	200	Maximum number of one batch
Join requests		
MaxPackageInterval	2 h	Maximum intervals between two packages of one end-device
MinPackageInterval	10 min	Minimum intervals between two packages of one end-device
Timeout	300 s	Timeout of each join request
Application packages by authorized end-devices and		
Application packages by authorized and unauthorized end-devices		
MaxPackageInterval	17 s	–
MinPackageInterval	13 s	–
Timeout	30 s	Timeout of each application package

transactions that are received within 2 s exceeds 200, a new block is also forced to be created.

Three kinds of experiments are conducted, i.e.,

1. **Experiment 1 (join requests):** 100–2,200 end-devices are emulated to continuously send join requests to LoRa gateways. Each LoRa gateway manages 25–550 end-devices. As shown in Fig. 5.8a, each end-device runs periodically with a randomly chosen interval between 10 min and 2 h. If a join accept message is received by end-device, the join procedure is considered successful. If no apply is received after 300 s, the join request is failed. The detailed processing time of join requests is recorded.

2. **Experiment 2 (application packages by authorized end-devices):** 100–2,000 end-devices are emulated in this experiment. Each LoRa gateway covers 25–400 end-devices equivalently. The join procedures of all end-devices are accomplished. Thus, end-devices are authorized to send application packages. Each end-device randomly chooses an interval between 13 s and 17 s to upload packages constantly. The network server replies each package with an acknowledgment to indicate that the uplink data are received successfully, as shown in Fig. 5.8b. If no acknowledgment is received by end-device for 30 s after it uploads, the transmission of application package is considered failed. End-device keeps sending after the time interval or any failure occurs. By running this experiment, the maximum possible performance of the prototype can be evaluated. The processing time, CPU utilization, and system throughput are recorded.

3. **Experiment 3 (application packages by authorized and unauthorized end-devices):** In this experiment, 100–2,000 end-devices are emulated in total. Four LoRa gateways share an equal number of end-devices. Among all the end-devices under the coverage of each LoRa gateway, half of them have finished

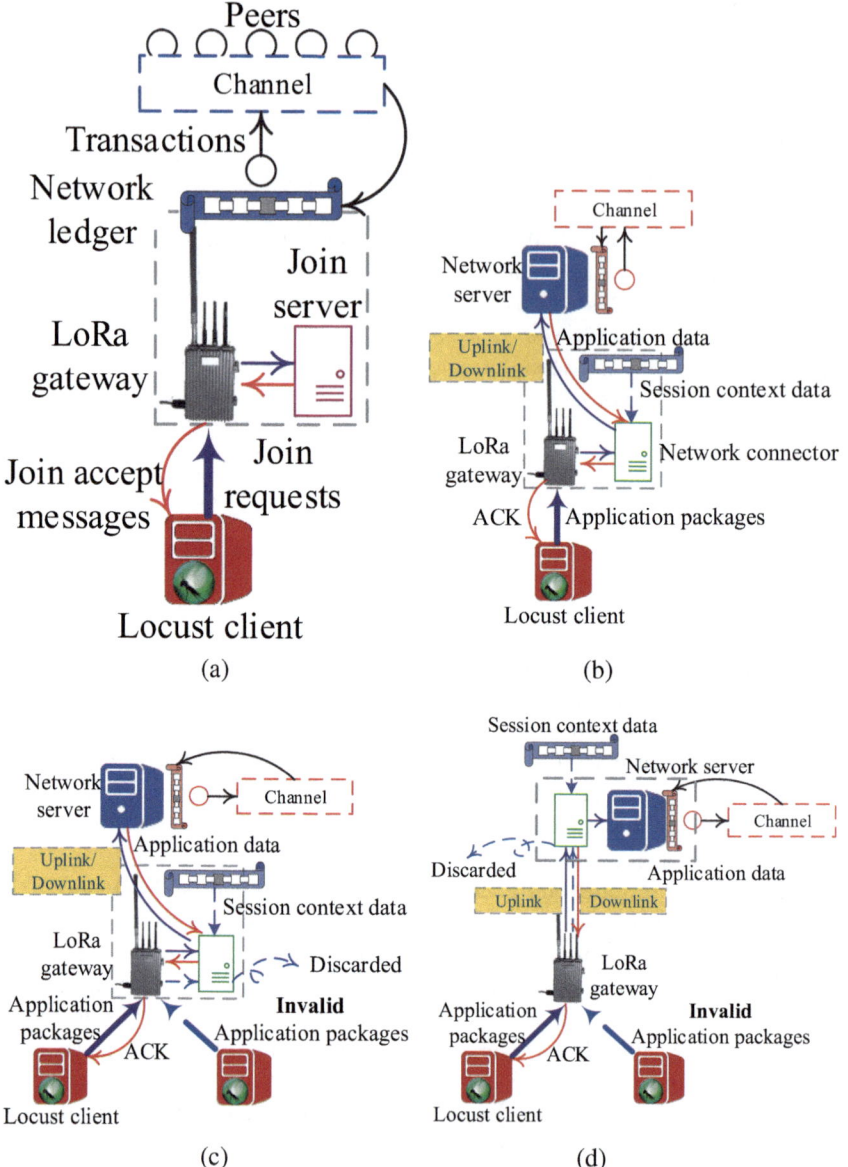

Fig. 5.8 Illustrations of the experiments on HyperLoRa prototype. (**a**) *Experiment 1*: join requests. (**b**) *Experiment 2*: application packages by authorized end-devices. (**c**) *Experiment 3*: application packages by authorized and unauthorized end-devices. (**d**) Comparison experiment with traditional LoRa system in *Experiment 3*

the join procedures, while the other half are considered unauthorized to the prototype. All the end-devices send application packages to LoRa gateways in the same way as *Experiment 2*. For those unauthorized end-devices, their packages cannot pass the MIC verification process in LoRa gateways and will be discarded directly, as shown in Fig. 5.8c. This experiment can simulate the situation that the system is under the application layer DoS attack. The system throughput, CPU utilization, and network bandwidth occupation are recorded. Moreover, a comparison experiment with traditional LoRa system approach is conducted. The application packages are sent to a traditional LoRa system where both the JS and NC are implemented in central cloud rather than in the LoRa gateways. Therefore, central cloud needs to process all valid and invalid application packages. A blockchain network is also established in the traditional LoRa system, but both $\mathcal{N}$ and $\mathcal{A}$ are maintained only by network servers, as illustrated in Fig. 5.8d. The comparison can show the improvement of performance brought by HyperLoRa.

5.6.3 Results and Analysis

Figure 5.9 illustrates the detailed distribution of processing time for join requests. Each bar represents the percentage of end-devices whose processing time falls within the value indicated by the top point for a specific number of end-devices. The processing time experiences a slight increase with the growing number of end-devices until reaching around 2,000. However, when the number of end-

Fig. 5.9 Distribution of processing time of join requests in *Experiment 1*

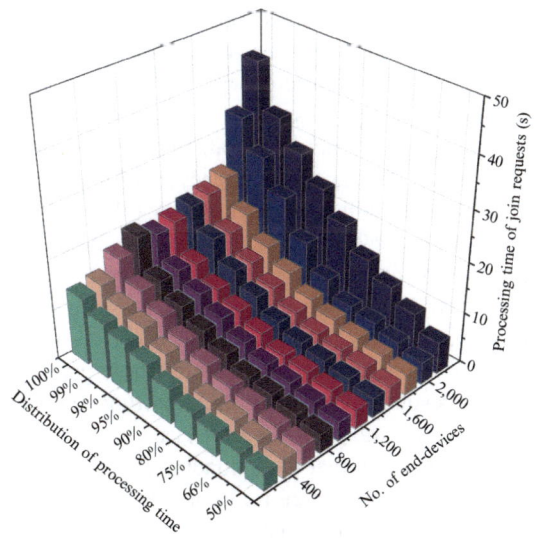

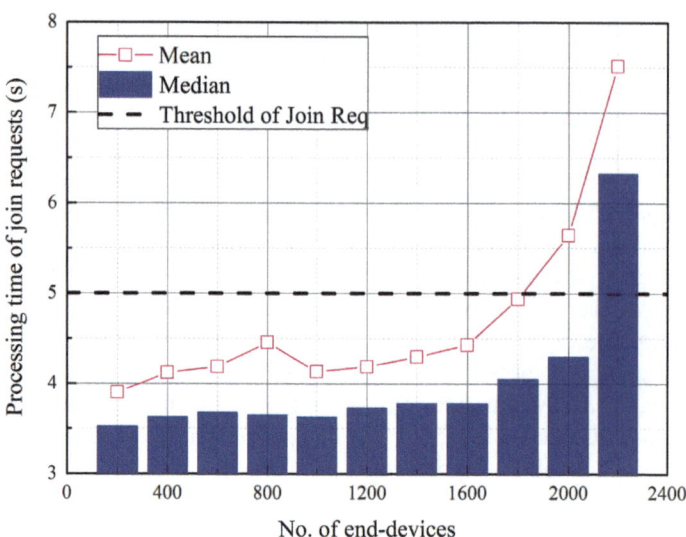

Fig. 5.10 Processing time statistics of join requests in *Experiment 1*

devices per LoRa gateway reaches approximately 500, there is a significant increase in processing time. The same trend is reflected in Fig. 5.10, which displays the statistical results of the processing time for join requests.

In Fig. 5.10, the black-dashed line represents the maximum allowable latency within which an end-device must receive the join accept message. It can be configured based on specific requirements (e.g., set to 5 s in *Experiment 1*). If the time exceeds this threshold, the join request is considered failed. The mean latency starts to exceed the threshold after the number of end-devices surpasses 2,000. These results highlight the maximum capability of handling join requests in the system; for a 5 s restriction and 1,800 end-devices, nearly 75% can successfully receive join accept messages.

Figure 5.11 depicts the statistical processing time of application packages in *Experiment 2*. The mean latency continues to increase as the number of end-devices rises until reaching around 1,100. This increase is due to the fact that more end-devices generate additional application data, which in turn requires more time to process. However, once the number exceeds 1,200, the processing time stabilizes. This stabilization occurs because the HyperLoRa system has reached its maximum load; there are no further resources available to handle additional application packages. Consequently, these packages are considered discarded.

This result indicates that the prototype can effectively manage a total of 1,200 end-devices, with uploading intervals of approximately 15 s and timeouts of 30 s. It is worth noting that the uploading frequency exceeds that of most sensing-based IoT applications. Therefore, in practice, the maximum number of end-devices that can be accommodated may still increase.

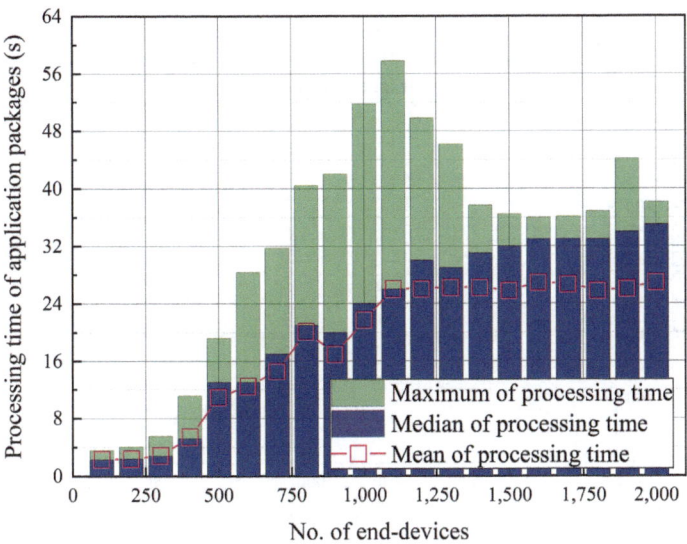

Fig. 5.11 Illustration of processing time of application packages in *Experiment 2*

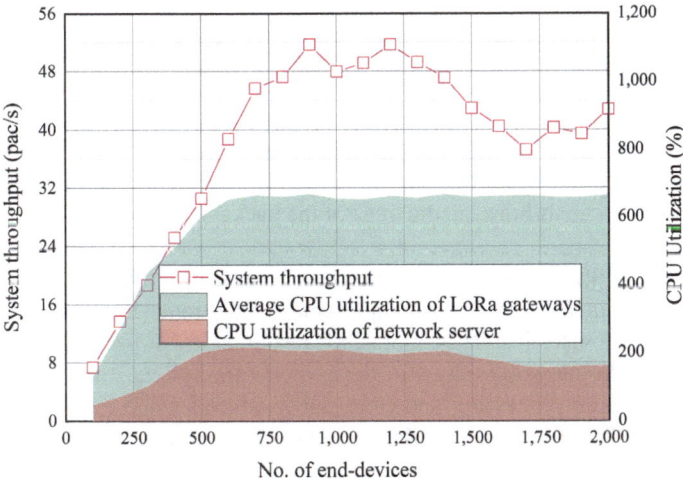

Fig. 5.12 Illustration of system throughput of application packages and CPU utilization of LoRa gateways and network server in *Experiment 2*

During the processing of valid application packages, the system throughput and the CPU utilization of both LoRa gateways and the network server are shown in Fig. 5.12. Initially, when there are fewer than 1,100 end-devices, the throughput keeps rising because the resources are not fully utilized. Concurrently, the CPU utilization is increasing. However, after the number of end-devices exceeds 1,200, the throughput stabilizes and even slightly declines. This trend is similar to what is

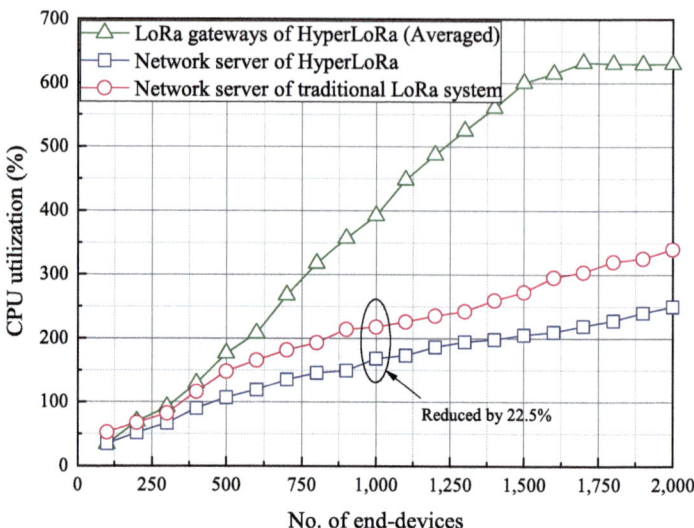

Fig. 5.13 Illustration of CPU utilization of LoRa gateways and network server with both valid and invalid application packages in *Experiment 3*

observed in Fig. 5.11, indicating that the system has reached a full-load condition. This is also reflected by the change in CPU utilization. Nearly 7 cores of the CPU in the LoRa gateways are occupied by the NC module, while the remaining core is used by the operating systems and the signal processing modules. In contrast, the CPU utilization of the network server remains low because the LoRa gateways with edge computing capabilities handle most of the package processing and verification tasks.

Figure 5.13 illustrates the CPU utilization of both LoRa gateways and the network server in *Experiment 3* compared to a traditional LoRa system. As the number of end-devices increases, more CPU resources are utilized in both the LoRa gateways and the network server. In the HyperLoRa system, the LoRa gateways handle all package processing and filter out invalid packages, leading to a 22% reduction in CPU utilization on the network server compared to the traditional LoRa system where the network server processes all packages. Not only CPU resources are spared, but also the bandwidth of transmission link between LoRa gateways and network server is saved, which is shown in Fig. 5.14. The filtering of invalid traffic by the LoRa gateways results in a 41.1% bandwidth saving when the number of end-devices reaches 1,000, compared to the traditional LoRa system. This bandwidth saving continues to increase linearly with the number of end-devices. The efficiency of resource utilization in the LoRa gateways allows the spared resources to be allocated to processing more valid packages, thus protecting the system from the adverse effects of invalid or malicious traffic. Figure 5.15 shows that the system throughput of HyperLoRa matches that of the traditional LoRa

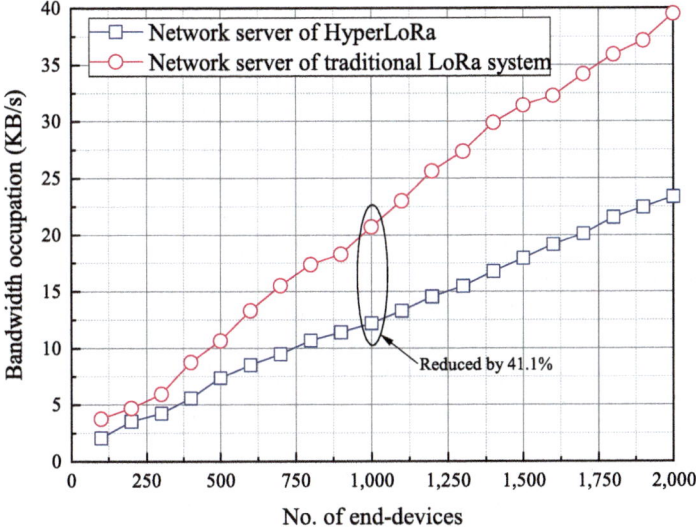

Fig. 5.14 Illustration of bandwidth occupation of transmission links between LoRa gateways and network server in *Experiment 3*

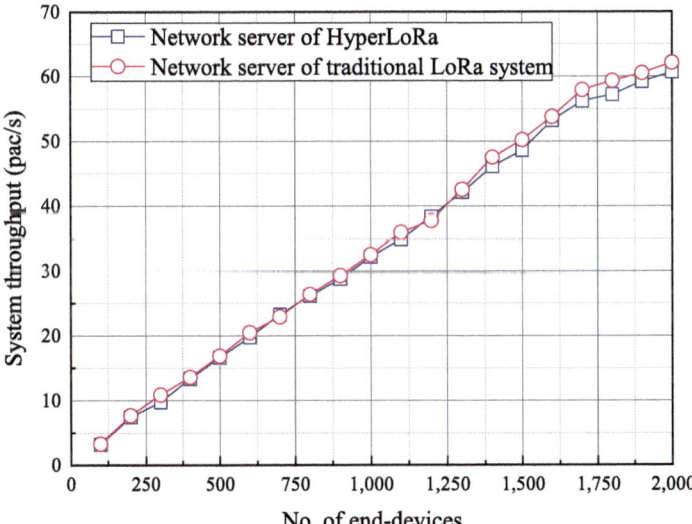

Fig. 5.15 Illustration of system throughput with both valid and invalid application packages in *Experiment 3*

system, indicating that HyperLoRa can achieve equivalent performance while using fewer resources. This efficiency is attributed to the optimized resource utilization in the LoRa gateways.

5.7   Summary

This chapter presents a prototype design for the LoRa system with edge computing, i.e., HyperLoRa, featuring the deployment of a blockchain with two ledgers to store different types of LoRa data. The HyperLoRa effectively utilizes the limited resources at LoRa gateways to maintain one ledger, while also migrating two network server functions to LoRa gateways to leverage their edge computing capabilities. It leverages edge computing and blockchain deployment to mitigate or prevent potential security risks. Furthermore, we have successfully implemented a prototype of HyperLoRa, demonstrating the feasibility of our provided design. Through experimentation, we have showcased the maximum performance levels achievable with HyperLoRa. For instance, it can handle nearly 1,350 join requests with a 5-second timeout and process 48 application packages per second with 1,000 end-devices uploading data at a 15-second interval. Moreover, compared to traditional LoRa systems, HyperLoRa enables the central cloud to conserve CPU utilization and bandwidth occupation without any reduction in system throughput.

References

1. S.M. Danish, M. Lestas, W. Asif, H.K. Qureshi, M. Rajarajan, A lightweight blockchain based two factor authentication mechanism for LoRaWAN join procedure, in *2019 IEEE International Conference on Communications Workshops (ICC Workshops)* (2019), pp. 1–6
2. J. Lin, Z. Shen, C. Miao, S. Liu, Using blockchain to build trusted LoRaWAN sharing server. Int. J. Crowd Sci. **1**(3), 270–280 (2017)
3. K.R. Ozyilmaz, A. Yurdakul, Designing a blockchain-based IoT with Ethereum, swarm, and LoRa: the software solution to create high availability with minimal security risks. IEEE Consum. Electron. Mag. **8**(2), 28–34 (2019)
4. Y. Liu, K. Wang, K. Qian, M. Du, S. Guo, Tornado: enabling blockchain in heterogeneous internet of things through a space-structured approach. IEEE Internet Things J. **7**(2), 1273–1286 (2020)
5. M.A. Ferrag, M. Derdour, M. Mukherjee, A. Derhab, L. Maglaras, H. Janicke, Blockchain technologies for the Internet of Things: research issues and challenges. IEEE Internet Things J. **6**(2), 2188–2204 (2019)
6. P. Yuan, K. Zheng, X. Xiong, K. Zhang, L. Lei, Performance modeling and analysis of a hyperledger-based system using GSPN. Comput. Commun. **153**, 117–124 (2020)
7. X. Wang, Y. Han, V.C.M. Leung, D. Niyato, X. Yan, X. Chen, Convergence of edge computing and deep learning: a comprehensive survey. IEEE Commun. Surv. Tutorials **22**(2), 869–904 (2020)
8. X. Wang, C. Wang, X. Li, V.C.M. Leung, T. Taleb, Federated deep reinforcement learning for internet of things with decentralized cooperative edge caching. IEEE Internet Things J. **7**(10), 9441–9455 (2020)
9. LoRaWAN™ backend interfaces 1.0 specification. LoRa Alliance (2017)
10. Z. Liu, Q. Zhou, L. Hou, R. Xu, K. Zheng, Design and implementation on a LoRa system with edge computing, in *2020 IEEE Wireless Communications and Networking Conference (WCNC)* (2020), pp. 1–6
11. LoRaWAN™ network server demonstration: gateway to server interface definition. Semtech Ltd (2015)

Chapter 6
Conclusion and Outlook

6.1 Conclusion

The IoT systems face significant challenges regarding data security, privacy, and trust. Although blockchain has demonstrated potential to address these issues, integrating blockchain with IoT remains challenging due to the computing, storage, and energy constraints of IoT devices. This monograph presents a comprehensive design for BIoT.

First, we present an architecture for BIoT that includes a multilayer blockchain network and a multi-ledger design, capable of handling various types of IoT data. We then detail the transaction and block designs and introduce a suitable consensus mechanism for BIoT. To enhance the efficiency of the BIoT system, we provide two optimization algorithms for block construction and consensus.

Furthermore, we introduce a transaction migration scheme that balances workloads across different edge servers. Our results indicate that this scheme can intelligently transfer transactions from hotspot areas to less busy areas. Considering the limited resources of IoT devices and the latency requirements of IoT applications, we introduce a new consensus algorithm, RAFT+. RAFT+ elects a leader based on resource occupation and communication channel status within the multilayer BIoT. The results show that RAFT+ can reduce the latency associated with new block consensus.

Finally, this monograph provides an in-depth look at the prototype implementation of a BIoT system using a LoRa network with edge computing. The system architecture, data flow, security mechanisms, and performance optimizations are detailed, showcasing the potential of integrating blockchain with IoT to create a secure, efficient, and scalable network. Experimental results demonstrate that the prototype enhances the security of the Long Range (LoRa) system while introducing minimal latency and workloads for LoRa devices and gateways. Moreover, the throughput of LoRa packages with blockchain integration still meets application requirements.

© The Author(s), under exclusive license to Springer Nature Switzerland AG 2024
L. Hou et al., *Blockchain-Based Internet of Things*, SpringerBriefs in Computer Science, https://doi.org/10.1007/978-3-031-70303-4_6

6.2 Open Challenges

When applying blockchain technology to the IoT, several significant challenges arise, impacting the feasibility and efficiency of such integration. Despite our efforts in exploration and research on BIoT, there are still significant issues and challenges that remain. These challenges include scalability, data storage, processing power and time, implementation complexity, costs, and so on.

- **Scalability**
  One of the key challenges in integrating blockchain with IoT is scalability. Classic blockchain systems face significant limitations in efficiently processing a high volume of transactions. IoT networks consist of numerous devices that generate massive amounts of data continuously. As the number of connected devices increases, the blockchain network must handle an increasing volume of transactions. Traditional blockchain systems struggle to meet this demand, resulting in low consensus efficiency and high latency. These issues are particularly challenging for IoT applications that require real-time data processing and low-latency responses, making scalability a significant obstacle in BIoT.
- **Data Storage**
  The decentralized nature of blockchain technology poses a substantial challenge in terms of data storage. IoT devices generate vast amounts of data that need to be securely stored and managed. Storing all of the data on the blockchain is impractical due to the continuously growing ledger. As more data are added, the blockchain ledger expands, requiring more storage space and computational resources. Usually IoT devices have limited storage capacity, and the continuously expanding blockchain ledger can exceed their capabilities, making the system unsustainable. Even with the help of edge computing servers, data storage still remains a severe challenge in BIoT. Effective data management solutions are needed to address these storage challenges without compromising the benefits of blockchain.
- **Processing Power and Time**
  The processing power and time required for blockchain operations present another significant challenge when integrating with IoT. Many blockchain systems use consensus mechanisms that demand substantial computational power and time to validate transactions. However, IoT devices often have limited computational capabilities and cannot perform resource-intensive operations, making many IoT applications impractical. As discussed in previous sections, new alternative consensus mechanisms need to be explored for BIoT. These mechanisms should require less computational power and time to make blockchain integration feasible for IoT devices.
- **Implementation Complexity**
  Blockchain and IoT technologies inherently have technical thresholds and considerable complexity. Combining the two further exacerbates these issues. Developing and deploying BIoT solutions require addressing various technical challenges, such as ensuring data integrity, managing decentralized trust, and

maintaining network security. Additionally, designing an effective decentralized system governance framework is complex. On the other hand, IoT typically has limited resources, including communication, computation, and storage, making the implementation of complex blockchain solutions an exceedingly daunting task.

- **Costs**

  Integrating blockchain with IoT can be costly due to the extensive resources required for data storage, processing, and maintaining the network. Storing large volumes of IoT-generated data on the blockchain leads to high storage costs. The computational power needed for various mechanisms in BIoT results in increased energy consumption and operational expenses. These costs can be prohibitive for companies to find cost-effective solutions for blockchain–IoT integration.

In summary, while blockchain technology offers significant potential to enhance the security, transparency, and efficiency of IoT networks, addressing challenges related to scalability, data storage, processing power and time, implementation complexity, and costs is crucial for feasible BIoT. Designing and implementing solutions that can manage large amounts of data, improve performance, ensure robust security, and establish effective governance frameworks will be key to overcoming these obstacles.

Index

© The Editor(s) (if applicable) and The Author(s), under exclusive license to
Springer Nature Switzerland AG 2024
L. Hou et al., *Blockchain-Based Internet of Things*, SpringerBriefs in Computer
Science, https://doi.org/10.1007/978-3-031-70303-4

GPSR Compliance

The European Union's (EU) General Product Safety Regulation (GPSR)
is a set of rules that requires consumer products to be safe and our
obligations to ensure this.

If you have any concerns about our products, you can contact us on
ProductSafety@springernature.com

In case Publisher is established outside the EU, the EU authorized
representative is:

Springer Nature Customer Service Center GmbH
Europaplatz 3
69115 Heidelberg, Germany

Batch number: 08304553

Printed by Printforce, the Netherlands